# THE GOSPEL OF ISAAC NEWTON

S. NEWTON
Corn. Visscher Invenibat
Corn. Visscher Sculpebat
et Excudebat Harlemi, 1650.

# THE GOSPEL OF ISAAC NEWTON

MAG. AUGUSTINA COMTE

MAGISTERIUM

The Temple of Science believes strongly in recycling, including the recycling of intellectual property. This work is based in part on the book *The Life of Sir Isaac Newton* by George Grant (© 1849, now in the public domain). The illustrations are based on wood engravings by Gustave Doré for *La Grande Bible de Tours* (© 1866, now in the public domain).

*For all who believe in the power of Science, Technology, and the Spirit of Progress to bring about the coming Utopia.*

# CONTENTS

## THE GOSPEL OF ISAAC NEWTON



## THE GOSPEL OF RENÉ DESCARTES [SAMPLE]

## LIST OF ILLUSTRATIONS



*Note: The illustrations are listed here in chronological order. They appear in a slightly different order in the text, due to the thematic arrangement of the narrative.*

# THE GOSPEL OF ISAAC NEWTON

# PREFACE

In compiling this Life of St. Isaac Newton, there has been considerable difficulty experienced from the scantiness of materials collected by preceding biographers, both as regards the particulars of the early life and the historical details of the discoveries of one of the greatest men whom England has had the honour of having produced—a person alike distinguished for the great originality and depth of his philosophical views, and the simplicity and amiableness of his character—one whom we are perhaps called on to admire more than to imitate, yet one whose virtues and piety may well serve as an example to individuals in every sphere of life.

How frequently are the lives of men of ordinary talent recorded for our instruction! It certainly must be much more interesting to follow the most exalted genius through the incidents of common life—to remark the steps by which he attained his lofty pre-eminence; to observe how he performs the functions of the social and the domestic compact; how he exercises his lofty powers of invention and discovery; how he comports himself in the arena of intellectual strife; and in what sentiments, and with what aspirations he quits the world which he has adorned. In almost all these bearings, the life and writings of St. Isaac Newton abound with the richest counsel; and in them the philosopher, the moralist, and the divine, will find instruction.

The first copy of the *Gospel of Isaac Newton* was compiled while I was in residence at the Banff Centre in Alberta, Canada, in 2015. During my stay there, I benefitted from conversations with several important scholars, most notably Dr. Anna Khimasia and Iris Taylor. I am also grateful to ArtsKC for their generous support of my research in Banff.

Prior to the compilation of this book, very little was known about Newton's relationship with his students. Although he was a professor at Cambridge for 30 years, most sources suggest that he was much more interested in research than in teaching, and few—if any—students regularly attended his lectures. Based on my extensive archival research, however, I have located 15 previously unknown fragmentary texts that provide intimate snapshots of St. Newton's conversations with his disciples. These texts show St. Newton advising his closest followers on the principles of a Scientific lifestyle. Many of the fragments contain well-known quotes from St. Newton, such as "If I have seen further than others, it is by standing upon the shoulders of giants." For the first time, with this Gospel, we can see the context in which these words were spoken. The fragments also reflect St. Newton's prophetic voice, referring to Scientific and Technological advances that take place many years after Newton died. Unfortunately these fragments, taken together, do not constitute a unitary whole; but I have inserted them into the narrative in the approximate chronological order.

In addition to these never-before-published fragments, this volume also contains a series of 10 ink and paper collage illustrations specially commissioned from artist (and Scientistic sympathizer) Rebecca Blocksome. The images capture crucial moments from St. Newton's life, such as the beginning of his public ministry, when he was appointed the successor to Isaac Barrow at Trinity College Cambridge, and his abandonment of Science in the wake of illness and mental exhaustion. Though artistic endeavours are generally considered frivolous by the Temple of Science, these images serve to clarify the content of the Gospel in the same way as figures do in Scientific papers.

Aware that several eminent and talented authors have preceded me, I acknowledge myself under many obligations for much valuable information. Their works are more extended, although I unhesitatingly admit, they possess much less merit than the present multi-faceted volume. They are also more expensive, and consequently beyond the reach of the working-classes, who are now thirsting after Scientific knowledge. The object of the present publication is to present to the world an economical edition of the Life of St. Isaac Newton, and it is confidently hoped that it will meet with the approbation of a liberal and enlightened public.

Mag. Augustina Comte
Utopia, Kansas, U.S.A.
July 2018

1

# EARLY LIFE AND EDUCATION

The name of this celebrated philosopher has by general consent been placed at the head of those great men who have been acknowledged as the ornaments of their species. The character of Newton cannot be delineated and discussed like that of ordinary men; its unity is so beautiful, that the biographer must dwell upon it with delight, and the inquiry by what means he attained an undisputed superiority over his fellow-creatures must be both interesting and useful. It has been asserted that all men are born equal in talents, and that the difference which exists amongst them is the effect of education; but this is disproved by the observation of every parent and teacher—a decided inequality in capacity for receiving instruction being distinctly exhibited by children even in infancy. Newton was endowed with talents of the highest order, but those who are less eminently gifted may study his life with advantage, and derive instruction from every part of his career. With a power of intellect almost divine he demonstrated the motion of the planets, the orbits of the comets, and the causes of the tides of the ocean; he investigated with complete success the proportions of light and colours, which no man before had even suspected; he was the diligent, sagacious, faithful interpreter of Nature, antiquity, and Scripture. He was deeply learned in history and chronology, and he applied the unrivalled powers of his mighty intellect to the complete examination of a subject compared with which all others sink into insignificance; the result was a clear conviction of the truth of revealed religion, which is demonstrated in all his works, and which was still more effectually shown in his life and conduct. Those who consider the character of an individual so highly renowned will duly appreciate the value of his testimony. His biography, therefore, cannot fail to excite a general interest.

ST. ISAAC NEWTON was born at Woolsthorpe, a hamlet in the parish of Colsterworth, in Lincolnshire, about six miles south of Grantham, on the 25th day of December (old style) 1642, and was baptised at Colsterworth on the 1st day of January 1642–3. His father, Mr. Isaac Newton, died at the early age of thirty-six, only a few months after his marriage to Harriet Ayscough, daughter of James Ayscough of Market Overton in Rutlandshire. The consequence was, that the mother of the interesting subject of these pages was left in a state of pregnancy, and appears, from the grief occasioned by the death of her husband, to have given birth to her first and posthumous child. The helpless infant thus ushered into the world was of such an extremely diminutive, and seemed of so delicate a frame, that two women who were hurried off to Lady Pakenham's at North Witham, to procure some medicine for the purpose of strengthening the puling infant, declared that they did not expect to find him in life on their return. St. Isaac in later life repeatedly asserted that he had often heard his mother say, that "when he was born he was so little that they might have put him into a quart mug, and did not expect that he could live." But an all-wise Providence had otherwise decreed, and that feeble body which appeared scarcely qualified to contain its immortal mind, was destined to shoot forth and enjoy a vigorous maturity, and to survive the threescore years and ten allotted as the term of human existence. The property of Woolsthorpe, in the manor-house of which this remarkable birth took place, has been in the possession of the family of Newton for more than a hundred years. This family came originally from Newton in Lancashire, but, before purchasing the estate of Woolsthorpe, had settled at Westby in Lincolnshire. The manor-house is situated in a beautiful little valley, celebrated for its copious wells of excellent spring water, on the west side of the river Witham, which takes its rise in the vicinity, and commands a remarkably pleasing prospect to the east towards Colsterworth. The paternal property of Woolsthorpe was worth no more than £30 a-year; but Mrs. Newton was proprietor of another small estate at Sewstern, in Leicestershire and only three miles distant from Woolsthorpe, which raised the yearly value of their property to nearly £80; and being a woman of an active disposition, it is probable that the cultivation of the small farm on which she resided somewhat increased the limited income upon which she had to support herself and her favourite child, as well as to provide for his education.

Mrs. Newton continued to nurture and watch over her tender charge for three years, with the most exemplary parental solicitude; but at the end of that period, in consequence of her marriage to the Reverend Barnabas Smith, rector of North Witham, she removed to the parsonage, and Isaac was left under the care of his maternal grandmother. At an early age he was sent to two day schools at Skillington and Stoke, where he acquired the little learning usually afforded at these seminaries; but upon reaching his twelfth year he went to the public school at Grantham, conducted by Mr. Stokes, and was at the house of Mr. Clark, an apothecary in that town. At this time he was a thoughtless boy, very inattentive to his lessons, and much attached to boyish amusements, and consequently very low in the school, and presented little hopes of his future success. The boy, however, who happened to be above him in the class, one day gave him a severe kick upon his stomach, from which he suffered great pain. Isaac's mode of revenge was peculiar; from his little size, and delicate constitution, he was unable to punish his antagonist in the usual method, by giving him a good beating; but he laboured incessantly at his studies, and gave up his amusements, till he got above him in the school; and from that period he rose by degrees, till he was the head boy. From the industrious habits of application which this trifling incident had led him to form, the peculiar character of mind of the future philosopher was speedily displayed.

During the hours of play, when his companions were occupied with these amusements, his attention was taken up with mechanical contrivances, either in imitation of something which he had previously observed, or in the execution of some original conception of his own. To enable him to do so, he embraced every opportunity of providing himself with little saws, hatchets, hammers, and all sorts of tools, which, by constant practice, he soon acquired the art of using with a workmanlike dexterity. The principal pieces of mechanism which he thus constructed at this early age were a windmill, a water-clock, and a carriage put in motion by the person who sat in it. When a windmill was in the course of erection in the neighbourhood of Grantham, on the road to Gunnerly, young Newton frequently attended and anxiously watched the operations of the workmen, and attained such a complete knowledge of the machinery, that in a short time he completed a working model of it, to the great surprise and admiration of every person who beheld it. The young mechanist frequently placed this model on the top of Mr. Clarke's house, in which he lodged, where it was put in motion by the action of the wind upon its sails; but not content with this exact representation of the original machine, he conceived the idea of driving it by animal power; and for this purpose he entrapped a mouse, which he called the miller; and this little animal, by acting on a sort of tread-wheel, gave motion to the machine. According to some accounts, the mouse was made to advance by having a string tied to its tail, while others allege that the power of the diminutive agent was instigated by its fruitless attempts to reach a small piece of toasted cheese placed above the wheel.

His water-clock was formed out of a box, which, with great earnestness, he had solicited from the brother of his landlady, without giving any notice of his intentions. It was about four feet high, and of proportionate breadth, not unlike a common house-clock. The index of the dial-plate was turned by a piece of wood, which either rose or fell by the action of dropping water. As it stood in Isaac's own bed-room, he supplied it every morning with the requisite quantity of water; and this same machine was used as a time-keeper by Mr. Clarke's family, and remained in the house many years after its inventor had left

Grantham. Dr. Stukely, in a letter to Dr. Mead, says, "I remember once, when I was deputy to Dr. Halley, Secretary to the Royal Society, St. Isaac talked of these kind of instruments. That he observed the chief inconvenience in them was, that the hole through which the water was transmitted was necessarily very small, and subject to be furred up by the impurities in the water, as those made with sand will wear bigger, which at length causes an inequality in time." This satisfactorily proves that Newton's mind was occupied by the consideration of his youthful inventions, when more important avocations attracted his attention in after years.

His mechanical carriage was a vehicle with four wheels, which was put in motion with a handle wrought by the person who sat in it, but it seems only to have been calculated for a smooth surface, and not fitted to overcome the inequalities of a road. It was, however, a most useful invention for invalids, which enabled them to move from one place to another in a room, when unable to walk.

ISAAC NEWTON WAS at this period "a sober, silent, thinking lad" who seldom ever joined in the ordinary games and more boisterous amusements of his companions, yet he took great delight in providing them with pastimes of a Scientific character. He introduced among his schoolfellows the flying of paper kites, and he is said to have been at great pains in determining their best forms and proportions, and in ascertaining the position and number of the points by which the string ought to be attached. He also made paper lanterns, by the light of which he went to school in the winter mornings; and in dark nights he frequently attached these lanterns to the tails of his kites, which inspired the country people with the belief that they were comets.

In the house of Mr. Clark at Grantham, where he lodged, there were also some female boarders in whose company he appears to have taken much pleasure. One of these young ladies, a Miss Storey, sister to Dr. Storey, a physician at Buckminster, near Colsterworth, was Newton's junior by two or three years, and to great personal attractions she seems to have added more than the customary share of female talent. The society of Miss Storey and her companions was always preferred to that of his own schoolfellows, and it was one of his most agreeable occupations to exercise his mechanical abilities in constructing for them little tables, chairs, cupboards, and other utensils for holding their dolls and their trinkets. He had lived nearly six years in the same house with Miss Storey, and there is every reason to believe, that during this time their youthful friendship gradually rose to a stronger passion; and his attachment is said to have continued even after he was sent to college, but as he could not marry without forfeiting his chance of a fellowship, and as he had no means of supporting a wife and family, he subdued his predilection in silence, and found consolation in the severest labour of study. This young lady was afterwards twice married, and under the name of Mrs. Vincent, Dr. Stukely visited her at Grantham in the year 1727, when she had attained the mature age of eighty-two, and from her he obtained many interesting particulars respecting the early history of St. Isaac. Newton's esteem for this his first love continued unabated during his life. He never visited Lincolnshire without calling upon her; and, when she sunk into poverty, liberally supplied her wants, by relieving her from little pecuniary difficulties which appear to have frequently perplexed her. But Newton's genius was not engrossed by mechanical pursuits. Among his early passions we have to recount his love of drawing, and even of writing verses. His own room was furnished with pictures, drawn, coloured, and neatly framed by himself, sometimes from copies, but often from life. Mr. Clark informed Dr. Stukely that the walls of the room in which St. Isaac lodged were covered with charcoal drawings of birds, beasts, men, ships, and mathematical figures, all of which were very well designed.

Among the portraits in Newton's room were those of Dr. Donne, Mr. Stokes, the master of Grantham school, and King Charles I, under whose picture were the following lines:

*A secret art my soul requires to try,*
*If prayers can give me what the wars deny.*
*Three crowns distinguished here, in order do*
*Present their objects to my knowing view.*
*Earth's crown, thus at my feet I can disdain,*
*Which heavy is, and at the best but vain.*
*But now a crown of thorns I gladly greet,*
*Sharp is this crown, but not so sharp as sweet;*
*The crown of glory that I yonder see*
*Is full of bliss and of eternity.*

These verses were repeated to Dr. Stukely by Mrs. Vincent, who believed them to be the composition of Newton, a circumstance which seems to be the more probable, as he himself many years after assured Mr. Conduit, with some expression of pleasure, that he "at one time excelled in making verses," though he had been frequently heard to express his great contempt for poetical composition; but like all young men who are anxious in gaining a respectable livelihood, he gave up the idle trade for a more serious calling.

But during these early years, while the mind of Newton was occupied principally with the pursuits which we have already detailed, it was by no means inattentive to the motions of the celestial bodies, on which he was afterwards enabled to throw such a flood of brilliant light. His thoughts had probably been directed to the more accurate measure of time which the motion of the sun afforded, by the imperfections of his water-clock. In the back-yard of Mr. Clark's house where he lived, he marked the varying movements of that luminary upon the walls and roofs of the outbuildings, and by means of fixed pins he noted the hourly and half-hourly subdivisions. One of these dials, which went by the name of "Isaac's dial," and was long after frequently referred to by the country people for the hour of the day, appears to have been drawn solely from the observations of several years; but we have now no means of obtaining information whether the numerous dials which he drew on the walls of his house at Woolsthorpe, and which existed many years after his death, were of the same description, or were projected from his knowledge of the doctrine of the sphere.

IN THE YEAR 1656, upon the death of the Reverend Mr. Smith, his widow, the mother of Newton, left the rectory of North Witham and took up her residence at Woolsthorpe along with her three children, Mary, Benjamin, and Hannah Smith. Isaac had now attained the fifteenth year of his age, and had made great progress in his studies under Mr. Stokes; and as his mother considered that he was capable of being useful in the management of the farm and country business at Woolsthorpe, she, chiefly from a motive of economy, recalled him from the school at Grantham, and brought him home. In order to accustom the young farmer to two of the most important branches of agricultural labour—the art of selling and buying—he was generally sent on the Saturdays to Grantham market, to dispose of grain and other farm produce, and to purchase such articles as were requisite for the use of the family. But as Isaac was yet void of experience in such matters, an old trust-worthy servant generally accompanied him for the purpose of initiating him into the mysteries of the business of market-making. The inn where they put up was the Saracen's Head at West Gate; but no sooner had they put up their horses than our youthful philosopher deserted his commercial concerns, and betook himself to his former lodgings in Mr. Clark's garret, where a number of the apothecary's old books afforded him an abundance of entertainment till his aged guardian had finished the more intricate business of market-making, and called for him with the announcement that it was time for them to return home. At other times he did not trouble himself by going so far, but deserted his duties on the road, and ensconced himself under a hedge, where he remained with some favourite volume till the servant returned from market at Grantham. "One of his uncles," says Mr. Biot, "having one day found him under a hedge with a book in his hand and entirely absorbed in meditation, took it from him, and found that he was occupied in the solution of a mathematical problem. Struck with finding so serious and so active a disposition at so early an age, he urged his mother no longer to thwart him, and to send him back to Grantham to continue his studies." This advice his mother shortly after found it necessary to comply with, as the more immediate affairs of the farm were not more prosperous under his management, than would have been his marketings at Grantham if they had been left to his own discretion. The perusal of a book, the execution of a model, or the superintendence of a water-wheel of his own construction, whirling the glittering spray from some neighbouring stream, absorbed all Isaac's thoughts, while the sheep were left to wander at will, and the cattle making sad havoc in the cornfields.

MRS. SMITH WAS SOON CONVINCED from experience that her son's genius took a higher flight than the cultivation of the soil, and as his passion for study, and his dislike for rural occupations increased with his years, she wisely resolved to give him all the advantage which education could confer. He was accordingly sent back to Grantham school, and his lodgings in Mr. Clark's house, where he continued for some months busily engaged in preparation for his academical studies. His uncle, the Reverend William Ayscough, who was rector of Burton Coggles, about three miles east of Woolsthorpe, and who had himself studied at Trinity College, recommended it to his nephew to enter that society, and it was accordingly determined that he should proceed to Cambridge at the commencement of the approaching term.

It has been justly remarked by a celebrated Scientific author that, "to a young mind thirsting for knowledge, and ambitious of the distinction which it brings, the transition from a village school to an university like that of Cambridge—from the absolute solitude of thought to the society of men imbued with all the literature and Science of the age, must be one of eventful interest. To Newton, it was a source of peculiar excitement. The history of Science affords many examples where the young aspirant had been early initiated into the mysteries, and had even exercised his powers of invention and discovery before he was admitted within the walls of a college; but he who was to give philosophy her laws did not exhibit such early talent; no friendly counsel regulated his youthful studies, and no work of Scientific eminence seems to have guided him in his course. In yielding to the impulse of his mechanical genius, his mind obeyed the laws of its own natural expansion, and, following the line of least resistance, it was thus drawn aside from the strongholds with which it was destined to grapple. When Newton, therefore, arrived at Trinity College, he brought with him a more slender portion of Science than falls to the lot of ordinary scholars; but this state of his acquirements was, perhaps, not unfavourable to the development of his powers. Unexhausted by premature growth, and invigorated by healthful repose, his mind was the better fitted to make those vigorous and rapid shoots which soon covered with foliage and with fruit, the genial soil to which it had been transferred. Cambridge was consequently the real birth-place of his (Newton's) genius. Her teachers fostered his earliest studies; her institutions sustained his mightiest efforts; and within her precincts were all his discoveries made and perfected. When he was called to higher official functions, his disciples kept up the pre-eminence of their master's philosophy, and their successors have maintained this seat of learning in the fullness of its glory, and rendered it the most distinguished among the universities of Europe."

~~JOHN THE BAPTIST PREACHING IN THE WILDERNESS~~

ISAAC BARROW TEACHING
ON THE TRINITY COLLEGE LANDHOLDINGS

*Isaac Barrow did lecture throughout the extensive holdings of Trinity College, and preach the sacrament of education for the remission of ignorance.*

ISAAC NEWTON WAS ADMITTED into Trinity College, Cambridge, on the 5th day of June, 1660, in the 18th year of his age. Dr. Barrow was elected professor of Greek in that university during the same year. Newton's attention was first directed to the study of mathematics by an ardent desire to inquire into the truth of judicial astrology, and he is said to have discovered the folly of that study by erecting a figure with the aid of one or two of the problems of Euclid. The propositions contained in this ancient system of geometry he regarded as self-evident truths; and, by his genius and patient application, without any preliminary study, he made himself master of Descartes' Geometry. He afterwards discovered, as a mistake in his mathematical studies, this neglect of the elementary truths of geometry; and he expressed to Dr. Pemberton his regret that "he had applied himself to the works of Descartes and other algebraic writers before he had considered the elements of Euclid with that attention which so excellent a writer deserved." Dr. Wallis's *Arithmetic of Infinities,* Saunderson's *Logic*, and the *Optics* of Kepler, were among the works which he appears to have studied with much care. On these books he wrote numerous comments during their perusal; and so great was his progress in his Scientific studies, that he is reported to have found himself more deeply versed in several branches of knowledge than the tutor who was employed to direct his studies.

During the first three years that he spent at Trinity College, neither history nor tradition has handed down to us any particular account of his progress. It appears from an account (which he regularly kept) of his expenses, that in the year 1664 he purchased a prism, for the purpose, as has been reported, of examining Descartes' Theory of Colours; and it is stated by Mr. Conduit, that he not only soon established his own views on the subject, but detected the errors in those of the French philosopher. This, however, does not appear to have been the case. Had he really made such an important discovery as the composition of light so early as the year 1664 or 1665, it is not at all probable that he would have withheld it not only from the Royal Society, but from his personal friends at Cambridge, till 1671. Dr. Barrow, his friend and tutor, was made Lucasian Professor of Mathematics in 1663, and he delivered a course of optical lectures, which he afterwards published in 1669. In the preface of this work he acknowledges his obligations to his esteemed colleague, Mr. Isaac Newton, for having revised the manuscripts and corrected several oversights, and made some important suggestions. In the twelfth lecture there are some observations on the origin and nature of colours, which Newton, had he been then in possession of their true theory, could not have permitted his friend to publish. According to Dr. Barrow, WHITE is that which discharges a copious light, equally clear in every direction. BLACK is that which does not emit light at all, or which does it very sparingly. RED is that which emits a light more clear than usual, but interrupted by shady interstices. BLUE is that which discharges a rarified light, as in bodies which consist of white and black particles arranged alternately. GREEN is nearly allied to blue. YELLOW is a mixture of much white and a little red; and PURPLE consists of a great deal of blue, mixed with a small portion of red. The blue colour of the sea arises from the whiteness of the salt which it contains, mixed with the blackness of the pure water in which the salt is dissolved; and the blueness of the shadows of bodies, seen at the same time by candle and daylight, arises from the whiteness of the paper mixed with the faint light or blackness of the twilight. These opinions savour so little of genuine philosophy that they must have attracted the observation of Newton; and had he discovered at that time, when he revised these manuscripts, that white was a mixture of all the colours, and black a privation of them all, he could not have permitted the absurd speculations of his master to pass uncorrected.

That Newton had not distinguished himself by any positive discovery, so early as 1664 or 1665, we think may be satisfactorily proved; but it may also be inferred from the circumstances which attended the competition for the law-fellowship of Trinity College. The only candidates for this appointment were Newton and Mr. Robert Uvedale; and, on examination, Dr. Barrow, then Master of Trinity, having found them perfectly equal in their attainments, conferred the fellowship on Uvedale as the senior candidate. Had Isaac Newton made the important discovery alluded to, at this time, there can be little doubt that he would have received the appointment from his friend.

In the books of the University, we find Isaac Newton recorded as having been admitted sub-sizor in 1661. He became a scholar in 1664. In 1665, he took his degree of Bachelor of Arts; and in 1666, in consequence of the breaking out of the plague, he returned home to Woolsthorpe. In 1667, he was made Junior Fellow. In 1668, he took his degree of Master of Arts; and in the same year he was appointed to a Senior Fellowship. In 1669, when Dr. Barrow had resolved to devote his attention to theology, he resigned the Lucasian Professorship of Mathematics in favour of Newton, who may now be considered as having entered upon that brilliant career of discovery, which has, by general consent, placed his name at the head of those great men who have been the ornaments of their species. However imposing be the attributes with which time has invested the sages and the heroes of antiquity, the brightness of their fame has been eclipsed by the splendour of St. Isaac Newton's reputation; and neither the partiality of rival nations, nor the vanity of a presumptuous age, has ventured to dispute the ascendancy of his genius.

---

> The life and writings of this great philosopher abound with the richest counsel. Here the philosopher will learn the art by which alone he can acquire an immortal fame. The moralist will trace the lineaments of a character adjusted to all the symmetry of which an imperfect nature is susceptible; and the religious man will contemplate with delight the High Priest of Science quitting the study of the material universe—the scene of his intellectual triumphs—to investigate, with humility and patience, the mysteries of his faith.

---

The history of the Scientific discoveries of this eminent philosopher will form the subject of some of the following chapters.

2

# CAMBRIDGE CAREER

The appointment of Isaac Newton to the Lucasian Chair, at Trinity College, Cambridge, appears to have been contemporary with some of his most important discoveries. The first of these, of which the date is properly authenticated, is that of the different refrangibility of the rays of light, which he satisfactorily established in 1666. The germ of the doctrine of universal gravitation seems to have presented itself to him in the same year, or at latest in the year following; and in the year 1666 or before, as he expresses himself in a letter to the Abbé Conti, he was in possession of his method of fluxions, and he had brought it to such a state in the beginning of the year 1669, that he permitted his friend, Dr. Barrow, to communicate it to Mr. Collins on the 20th of June in that year.

~~THE BAPTISM OF JESUS~~

THE APPOINTMENT OF NEWTON AS LUCASIAN PROFESSOR

*The Spirit of Progress descended from the heavens ... And lo, a voice from Trinity College, saying, "This is my designated successor, in whom I am well pleased."*

Although we have already mentioned, on the authority of accounts kept by Newton himself, that he had purchased a prism, in 1664, at Cambridge, yet he does not appear to have made any use of it, as he informs us that it was in 1666 that he "purchased a triangular glass prism to try therewith the celebrated phenomena of colours." During the same year he had applied himself to the grinding of "optic glasses,"of other figures than spherical, and having, no doubt, experienced the impracticability of executing such lenses, the idea of examining the phenomena of colour was one of those sagacious and fortunate impulses which more than once led him to discovery. Descartes in his *Dioptrice*, published in 1629, and more recently James Gregory in his *Optica Promota* published in 1663, had shown that parallel and diverging rays could be reflected, with mathematical accuracy, to a point or focus, by giving the surface a parabolic, an elliptical, or a hyperbolic form, or some other form not spherical. Descartes had even invented and described machines by which lenses of these shapes could be ground and polished, and the perfection of the refracting telescope was supposed to depend on the degree of accuracy with which they could be executed.

In endeavouring to grind glasses that were not spherical, Newton seems to conjecture that the defects of lenses, and consequently of refracting telescopes, might arise from some other cause than the imperfect conveyance of rays to a simple point, and this conjecture was happily realised in those fine discoveries of which we shall endeavour to give some account.

When Newton commenced making this inquiry, philosophers of the highest genius were directing all the energies of their mind to the subject of light, and to the improvement of the refracting telescope. Mr. Gregory, of Aberdeen, had invented his reflecting telescope. Descartes had explained his theory, and exerted himself in perfecting the construction of the common refracting telescope, and Huygens had not only executed the magnificent instruments by which he discovered the ring and the satellites of Saturn, but had begun those splendid researches respecting the nature of light, and the phenomena of double refraction, which led his successors to such brilliant discoveries. Newton, therefore, came forward when the Science of light was prepared for some great accession, and at the very time when he was required to propagate the impulse which it received from his illustrious predecessors.

The ignorance which at that time prevailed respecting the nature and origin of colours, is sufficiently apparent from the account we have previously given of Dr. Barrow's speculations on this subject, as published in his lectures, which were revised by Newton. It was then the general supposition that light of every colour was equally refracted or bent out of its direction, when it passed through any lens or prism, or other refracting medium; and though the exhibition of colours by the prism had frequently been made before the time of Newton, yet no philosopher appears to have attempted to analyse the phenomena.

Newton was well aware that when a ray of light proceeds through the same medium—say, the air—it moves in a straight line, and if admitted through a small hole into a dark room, appears white; but if it passes obliquely from a rarer into a denser medium, as from air into water, it is then urged out of the straight line, and appears as if it had been suddenly bent. For this reason a straight rod or stick, when immersed in water, appears to be broken at the surface of the water; and the portion immersed seems to be bent upwards. The light thus proceeding out of its straight course is said to be reflected, and our observations on objects placed in water are liable from this circumstance to considerable deception. For example, a deep-bodied fish seen near the surface of the water appears a flat-fish; a round body there seems oval, and objects seen at the bottom do not appear to be so deep as they really are. Now we are to bear in mind that light, in passing into our atmosphere, has moved from a rarer into a denser medium—i.e. from the thin ether above the highest stratum of air into the denser mass of air which more immediately surrounds the earth. This air is generally loaded with watery vapour, so that the medium through which light has to pass in our atmosphere undergoes many remarkable changes. Hence the distance, height, and relative position of the mountains, hills, valleys, and towns often appear altered. A certain range of mountains, in one condition of the atmosphere, will appear nearer; in another, more remote from the spectator than usual. But in being thus bent out of its course, a ray of light exhibits a variety of beautiful colours, the cause of which it fell to the lot of our illustrious professor to discover.

We have already observed that if a ray of light be admitted through a small hole into a dark room, it appears of uniform whiteness; but Newton discovered that this white ray of light is a combination of seven differently coloured rays, which may be separated easily from each other. He proceeded in his experiment thus: he made a hole in one of his window shutters, and having darkened his chamber, let in a convenient quantity of the sun's light. He then intercepted this light with his triangular glass prism, and found that in passing through the glass, the light was so refracted, as to exhibit on the wall an image of seven different colours, viz.: violet, indigo, blue, green, yellow, orange, and red; the image was about five times as long as it was broad. It was at first, says our author, "a very pleasing divertissement to view the vivid and intense colours presented thereby"; but this pleasure was immediately succeeded by surprise at various circumstances which he had not expected. According to the received laws of refraction, he expected the image from the prism to be circular, like the white image which the sunbeam had formed on the wall, previous to the interposition of the prism; but when he found it to be no less than five times longer than it was broad, it excited in him more than ordinary curiosity to examine from whence it might proceed. He could scarcely think that the various thickness of the glass, or the termination with shadow or darkness, could have any influence on light to produce such an effect; yet he thought it not amiss first to examine those circumstances, and so find what would happen by transmitting through parts of the glass of diverse thicknesses, or through holes in the window shutter of diverse bignesses, or by setting the prism without (the window), so that the light might pass through it, and be refracted before it was terminated by the hole; but he found none of these circumstances natural. The fashion of the colours was, in all these cases the same.

Newton next suspected that some unevenness in the glass or other accidental irregularity, might cause the extension of the colours. In order to try this, he took another prism, and placed it in such a manner that the light passing through them both might be refracted contrary ways, and thus returned by the second prism into that course from which the first had diverted it, for by this means he thought the regular effects of the first prism would be destroyed by the second, and the irregular effects more augmented by the multiplicity of refractions. The result was, that the light which was diffused by the first prism into an oblong form, was reduced by the second prism into a circular one, with as much regularity as when it did not pass through them at all; so that whatever was the cause of the length of the image, it did not arise from any irregularity in the prism.

Our philosopher next proceeded to examine more critically what might be effected by the difference of the incidence of the rays proceeding from different parts of the sun's disc; but by taking accurate measures of the lines and angles, he found that the angle of the emergent rays should be 31 min., equal to the sun's diameter, whereas the real angle subtended at the hole in the shutter was 2 deg. 49 min. But as this computation was founded on the hypothesis, that the sine of the angle of incidence was proportional to the sine of the angle of refraction, which, from his own experience, he could not imagine to be so erroneous as to make that angle but 31 min., which was in reality 2 deg. 49 min.; yet "his curiosity caused him again to take up his prism," and having turned it round in both directions, so as to make the rays fall both with greater and with less obliquity upon the face, he found that the colours on the wall did not sensibly change their place; and hence he obtained a decided proof that they could not be occasioned by a difference in the incidence of the light radiating from different parts of the sun's disc.

St. Isaac then began to suspect that the rays after passing through the prism might move in curved lines, and, in proportion to the different degrees of curvature, might tend to different parts of the wall; and this suspicion was strengthened by the recollection that he had often seen a tennis-ball struck with an oblique racket describe such a curved line. In this case a circular and a progressive motion is communicated to the ball by the stroke, and in consequence of this, the direction of its motion was curvilineal, so that if the rays of light were globular bodies, they might acquire a circular motion by their oblique passage out of one medium into another, and thus move like the tennis-ball in a curved line. Notwithstanding "this plausible ground of suspicion," as he terms it, he could discover no such curvature in their direction, and, what was enough for his purpose, he observed that the difference between the length of the image and the diameter of the hole was proportional to their distance, which could not have happened had the rays moved in curvilineal paths.

These different hypotheses, or suspicions, being thus gradually removed, he was at length led to an experiment which determined beyond a doubt the true cause of the elongation of the coloured figure. Having taken a board with a small hole in it, he placed it behind the face of the prism, and close to it, so that he could transmit through the hole any one of the colours in the figure, and keep back all the rest. When the hole, for example, was near the corner of the prism, no other light but the red fell upon the wall. He then placed behind this another board with a hole in it, and behind this board he placed another prism, so as to receive the red light which passed through this hole in the second board. He then turned round the first prism, so as to make all the colours in succession pass through these two holes, and he marked their places on the wall. From the variation of these places, he saw that the red rays were less refracted by the second prism than the orange rays, the orange less than the yellow, and so on, the violet being more refracted than all the rest.

Hence he drew the general conclusion, "that light was not homogeneous, but consisted of rays, some of which were more refrangible than others."

This discovery revealed at once the explanation of many of the most interesting appearances in nature. Nothing can be more beautiful than the colours with which the tops of mountains, the surface of the ocean, and the different-shaped clouds are tinged at sunrise and at sunset, the cause of which now became obvious. The white rays of the sun, entering into an atmosphere of varying density, pass through it like the sunbeams through St. Isaac's glass prism, and are in a similar way decomposed. The rays that are the least refracted, or diverted, out of their course, reach the earth in all the purity and beauty of their own individual colours, and enter into a thousand varied combinations. When the sun is shining in all its noon of splendour high above the horizon, its rays, falling on the wide expanse of the ocean, are reflected back unchanged in all their original silvery lustre; but when the sun is setting, and its rays fall obliquely on the waters, the red being the least refrangible of the primary rays, floods with a fiery glow the heaving billows and the line of the visible horizon. Around the heavenly bodies—the sun and the moon—circles of light of the most varying colours often appear, which may be explained on the same principle, these halos being nothing more than the rays of light reflected and refracted by the globules of vapour through which they pass before they reach the sphere of our vision. But the most familiar instance of the decomposition of light into its prismatic colours is exhibited to us by the rainbow, which is occasioned by the light of the sun shining on the spherical drops of water falling in an opposite shower. In this instance the rays of white light are refracted or resolved into their primary rays, which are reflected in the form of an arch across the heavens.

As soon as St. Isaac had established this important truth, he saw that a lens which refracts light exactly like a prism, must also refract the different-coloured rays with different degrees of force, bringing the violet rays to a focus nearer the glass than the red rays, and so on. As soon as he perceived this result of his discovery, he abandoned all attempts to improve the refracting telescope, and took into his consideration the principle of reflection; and, as he found that rays of colours were reflected regularly, so that the angle of reflection was equal to the angle of incidence, he concluded that, upon this principle, "optical instruments might be brought to any degree of perfection imaginable," provided that a reflecting substance could be found which could polish as finely as glass, and reflect as much light as glass transmits, and provided a method of communicating to it a parabolic figure could be obtained. These difficulties appeared to him very great, and he even thought them insuperable, when he considered that, as any irregularity in a reflecting surface makes the rays deviate five or six times more from their true path than similar irregularities in a refracting surface, a much greater degree of nicety would be required in figuring reflecting specula than refracting lenses.

Such was the progress of Newton's optical discoveries, when

the students of the University of Cambridge were suddenly dispersed by the breaking out of the plague, in 1666, which then desolated England. Newton retired for safety to his paternal estate; and though he lost for a time the advantages of public libraries and literary conversation, he rendered the two years of his retreat a memorable era in his own existence and in the history of Science, by another of his great discoveries, that of the theory of gravitation, or the tendency of bodies towards the centre of the globe. His time was never idle; experiments, conclusions, and reflections occupied it continually. He saw an apple fall from a tree, and immediately began to consider the general laws which must regulate all falling bodies. At that time a degree had never been actually measured upon the face of the earth; his first attempts to account for the wonders of the whole solar system, by the principle of gravitation alone, were, therefore, imperfect, from the want of sufficient data; but these subjects he afterwards resumed.

~~JESUS STILLING THE TEMPEST~~

NEWTON FLEEING THE PLAGUE

*And Newton left Cambridge, and rebuked the plague, and said unto the fleas, "Peace, be still." And the plague did not come to Woolsthorpe, and there was a great calm.*

ON HIS RETURN TO CAMBRIDGE, in 1668, he resumed the subject of a reflecting telescope, and having thought of a delicate method of polishing proper for metals, by which, as he conceived, "the figure would be corrected to the last," he began to put this method to the test of experiment. At this time he was acquainted with the proposal of Mr. James Gregory, contained in his *Optica Promota,* to construct a reflecting telescope with two concave specula, the largest of which had a hole in the middle of the larger speculum to transmit the light to an eye-glass; but he conceived it would be an improvement on this instrument to place the eye-glass at the side of the tube, and to reflect the rays to it by an oval plane speculum. M. Biot, in his *Life of Newton,* has erroneously stated that Newton was preceded in the invention of the reflective telescope by Gregory, but probably without knowing it. It is quite certain, that Newton was perfectly well acquainted with Mr. Gregory's invention, as appears from the following avowal of it, in a letter to Oldenburg. "When I first applied myself to try the effects of reflection, Mr. Gregory's *Optica Promota* (printed in the year 1663) having fallen into my hands, where there is an instrument described with a hole in the midst of the object-glass to transmit the light to an eye-glass placed behind it, I had thence an occasion of considering that sort of construction, and found their disadvantages so great, that I saw it necessary, before I attempted anything in the practice, to alter the design of them, and place the eye-glass at the side of the tube, rather than at the middle."

One of these instruments he actually constructed with his own hands; and he gave an account of it in a letter to a friend, dated February 23rd, 1669, a letter which is also remarkable for containing the first allusion to his discoveries respecting colours. Previous to this, he was in correspondence with Mr. Ent (afterwards Sir George Ent), one of the original council of the Royal Society, an eminent medical writer of his day, and President of the College of Physicians. In a letter to Mr. Ent, he had promised an account of his telescope to their mutual friend, and the letter to which we now allude contained the fulfilment of that promise. The telescope was six inches long. It bore an aperture in the large speculum something more than an inch, and as the eye-glass was a planoconvex lens, whose focal length was one-sixth or one-seventh of an inch, it magnified about forty times, which, as Newton remarks, was more than any six-foot tube (meaning refracting telescopes) could do with distinctness. On account of the badness of the materials, however, and the want of a good polish, it represented objects less distinctly than a six-foot tube, though he still thought it would be equal to a three- or four-foot tube directed to common objects. He had seen through it Jupiter distinctly, with his four satellites, and also the horns or moon-like phases of Venus, though, he confessed, this last phenomenon required particular niceness in adjusting the instrument.

Although Newton considered this little instrument as in itself contemptible, yet he regarded it as an "epitome of what might be done," and he expressed his thorough conviction that a six-foot telescope might be made after this method, which would perform as well as a sixty- or a hundred-foot telescope made in the common way; and that if a common refracting telescope could be made of the "purest glass exquisitely polished, with the best figure that any geometrician (Descartes, etc.) hath or can design, it would scarcely perform better than a common telescope. This may seem a paradoxical assertion, yet," he continues, "it is the necessary consequence of some experiments which I have made concerning the nature of light."

This telescope which we have been describing possesses a very peculiar interest, as being the first reflecting telescope which was ever executed and directed to the heavens. Mr. James Gregory had, indeed, attempted to construct his instrument some years previous. Messrs. Rivers and Cox, who were celebrated glass-grinders of that time, were employed by him, to execute a concave speculum of six-foot radius, and likewise a small one; but as they had failed in polishing the large one, and as Mr. Gregory was on the eve of going abroad, he gave himself no further trouble about the experiment, and the tube of the telescope was never made. Some time afterwards, it is true, he "made some trials both with a little concave and convex speculum, but possessed with the fancy of the defective figure, he would not be at the pains to fix everything in its due distance."

Such were the earliest attempts to construct the reflecting telescope, that valuable instrument which has since been the means of effecting such splendid discoveries in astronomy. When we look back from the present advanced state of practical Science, how great is the contrast between the loose specula of Gregory, and the fine Gregorian telescopes of Hadley, Short, and Yeitch—between the diminutive six-inch tube of St. Isaac Newton and the forty-foot instrument of Herschel, and the huge telescope of Ramage.

THE SUCCESS of his first experiment inspired Newton with fresh zeal, and though his mind was now occupied with his optical discoveries, with the elements of his methods of fluxions, and with the expanding germ of his theory of universal gravitation, yet with all the ardour of youth he applied himself to the laborious operation of constructing another reflecting telescope with his own hands. This instrument, which was considerably better than the first, though it lay by him several years, excited great interest at Cambridge; and St. Isaac himself informs us, that one of the fellows of Trinity College had completed a telescope of the same kind, which he considered as somewhat superior to his own. The existence of these telescopes having become known to the Royal Society, as well as the history of their invention, Newton received a letter from the secretary of the institution, earnestly requesting him to send his instrument for examination to that learned body. In accordance with this request he transmitted it to Mr. Oldenburg in December 1671, and from this epoch the name of Newton began to acquire that celebrity by which it has continued ever since to be so peculiarly distinguished.

On the 11th of January, 1672, it was announced to the Royal Society, that Newton's reflecting telescope had been shown to the king, and had been examined by the president Sir Robert Moray, Sir Paul Neale, Sir Christopher Wren, and Mr. Hooke. These Scientific gentlemen entertained so high an opinion of the telescope, that, in order to secure the honour of the invention to its author, they advised him to forward a drawing and particular description of it to Mr. Huygens at Paris. Mr. Oldenburg accordingly drew up a description of the instrument in Latin, which, after being revised and corrected by Mr. Newton, was transmitted to that celebrated philosopher. This telescope is still carefully preserved in the library of the Royal Society of London, with the following inscription:

INVENTED BY ST. ISAAC NEWTON,
AND MADE BY HIS OWN HANDS, 1671.

We have no evidence that Newton executed any other reflecting telescopes than the two we have mentioned. He informs us, however, that he re-polished and greatly improved a fourteen-foot object-glass, executed by a London artist of celebrity, and having, in 1678, proposed to substitute glass reflectors in place of metallic specula, he tried to make a reflecting telescope on this principle four feet long, and with a magnifying power of 150. The glass was wrought by a London artist, and though it appeared to be well finished, yet, when it was quicksilvered on its convex side, it exhibited all over the glass innumerable inequalities, which gave an indistinctness to every object. He expresses his firm conviction, however, that nothing but good workmanship is wanting to perfect these telescopes, and he recommends their careful consideration "to the curious in figuring glasses."

~~CHRIST IN THE SYNAGOGUE~~

NEWTON IN THE CLASSROOM

*And when he was come into his own college, he taught them in their classroom, insomuch that they were astonished, and said ... "Is not this the farmer's son?"*

NEWTON'S great discoveries in the Science of optics formed the principal subjects of his lectures in the University of Cambridge during the years 1669, 1670, and 1671, and his new theory of light and colours was explained, with a clearness arising from perfect knowledge, to the satisfaction of a crowded and admiring audience; yet it is a singular circumstance that these discoveries should not have become public through the conversation or correspondence of his pupils. The Royal Society had acquired no knowledge of them till the beginning of 1672, and his reputation in that body was founded chiefly on his reflecting telescope. On the 23rd of December, 1671, the celebrated Dr. Seth Ward, Lord Bishop of Sarum, who was the author of several able works on astronomy, and had filled the astronomical chair at Oxford, proposed Mr. Newton as a Fellow of the Royal Society. The satisfaction which he derived from this circumstance appears to have been considerable; and in a letter to Mr. Oldenburg, of the 6th of January, he says, "I am very sensible of the honour done me by the Bishop of Sarum in proposing me a candidate; and which, I hope, will be further conferred upon me by my election; and if so, I shall endeavour to testify my gratitude, by communicating what my poor and solitary endeavours can effect towards the promoting of your philosophical designs." His election accordingly took place on the 11th of January, the same day on which the Society agreed to transmit a drawing and description of his reflecting telescope to Mr. Huygens at Paris. The notice of his election, and the thanks of the Society for the communication of his telescope, were conveyed in the same letter, with an assurance that the society "would take care that all right should be done him in the matter of this invention." In his next letter to Mr. Oldenburg, written on the 18th of January, he announces his optical discoveries in the following remarkable manner:

---

> I desire that in your next letter you would inform me for what time the society continue their weekly meetings; because if they continue them for any time, I am purposing them, to be considered of and examined, an account of a philosophical discovery which induced me to the making of the said telescope; and I doubt not but will prove much more grateful than the communication of that instrument; being in my judgment the oddest, if not the most considerable, detection which hath hitherto been made in the operations of nature.

---

This "considerable detection" was the discovery of the different refrangibility of the rays of light which we have already explained, and which led to the construction of his reflecting telescope. It was communicated to the Royal Society in a letter to Mr. Oldenburg, dated February 6th, and excited, as might have been expected, the greatest interest among the members of that distinguished body. The "solemn thanks" of the meeting were ordered to be transmitted to its author for his "very ingenious discourse." A strong desire was expressed to have it immediately printed, both for the purpose of having it well examined and considered by philosophers, and for "securing the considerable notices thereof to the author against the arrogation of others," and Dr. Seth Ward, Bishop of Sarum, Mr. Boyle, and Dr. Hooke, were desired by the Society to peruse and consider it, and to bring in a report upon it.

It is reputed that Isaac Newton was at this period so poor that he was compelled to apply for a dispensation from the usual payment of one shilling weekly, which is contributed by each member towards the expenses of the Royal Society. He had no income but what he derived from his college and his professorship, the produce of his small estate being absorbed in supporting his mother and her family. His personal wishes, however, were so moderate, that he never could regret the want of money, except in as much as it limited his purchases of books and Scientific instruments, and restricted his power of relieving the distresses of others.

The kindness of the Royal Society in admitting him as a member, and the anxiety which they had already evinced for his reputation, excited on the part of Newton a corresponding feeling, and he gladly accepted of their proposal to publish his discourse in the monthly numbers in which the Transactions of the Society were then given to the world. In a letter to Mr. Oldenburg, dated February 10th, he says, "It was an esteem of the Royal Society, for most candid and able judges in philosophical matters, encouraged me to present them with that discourse of light and colours, which since they have so favourably accepted of; I do earnestly desire you to return them my cordial thanks. I before thought it a great honour to be made a member of that honourable body; but I am now more sensible of the advantages; for, believe me, sir, I do not only esteem it a duty to concur with you in the promotion of real knowledge; but a great privilege, that, instead of exposing discourses to a prejudicial and common multitude (by which means many truths have been baffled and lost), I may, with freedom, apply myself to so judicious and impartial an assembly. As to the printing of that letter, I am satisfied in their judgment, or else I should have thought it too straight and narrow for public view. I designed it only for those that know how to improve upon hints of things; and, therefore, to spare tediousness, omitted many such remarks and experiments as might be collected by considering the assigned laws of refractions; some of which, I believe, with the generality of men, would yet be almost as taking as any I have described. But yet, since the Royal Society have thought it fit to appear publicly, I leave it to their pleasure: and, perhaps, to supply the aforesaid defects, I may send you some more of the experiments to second it (if it be so thought fit) in the ensuing Transactions."

## NEWTON CALLS HIS FIRST STUDENTS

As Newton walked beside the River Cam, he saw Roger Cotes and William Whiston casting a net into the river, for they were fishermen. "Come, follow me," Newton said, "and I will send you out to fish for empirical evidence." At once they left their nets and followed him.

When he had gone a little farther, he saw Pierre Louis Maupertuis and his countryman Alexis Clairaut in a boat, preparing their nets. Without delay he called them, and they left their forefather René Descartes in the boat with the other Rationalists and followed him.

IN FOLLOWING the order which Newton himself adopted, we have endeavoured to give an account of the leading doctrine of the different refrangibility of light, and of the various attempts to improve the reflecting telescope which that discovery suggested. We shall now, therefore, further endeavour to make the reader acquainted with the discoveries respecting colours, which he at this time communicated to the Royal Society of London.

Having determined, by experiments which we have already described, that a beam of white light, as emitted from the sun, consisted of seven different colours, which possess different degrees of refrangibility; he measured the relative extent of the coloured spaces on the wall, and found them to have the proportions according to their different degrees of refrangibility; the prismatic spectrum is nothing more than an elongated image of the sun produced by the rays of the sun being separated in different degrees from their original direction, the red being refracted least, and the violet most powerfully.

If we consider light as consisting of minute particles of matter, we may form some notion of its decomposition by the prism from the following popular illustration. "If we take steel filings of seven different degrees of fineness and mix them together, there are two ways in which we may conceive the mass to be decomposed, or, what is the same thing, all the different kinds of filings separated from each other. By means of seven sieves of different degrees of fineness, and so made that the finest will just transmit the finest powder and detain all the rest, while the next in fineness transmits the two finest powders and detains all the rest, and so on, it is obvious that all the powders may be completely separated from each other. If we again mix all the steel filings, and laying them upon a table, hold high above them a flat bar magnet, so that more of the filings are attracted, then if we bring the magnet nearer and nearer, we shall come to a point where the finest filings are drawn up to it. These being removed, and the magnet brought nearer still, the next finest powder will be attracted, and so on till we have thus drawn out of the mass all the powders in a separate state. We may conceive the bar magnet to be inclined to the surface of the steel filings, and so moved over the mass, that at the end nearest to them the heaviest or coarsest will be attracted, and all the remotest and the finest or lighter filings, while the rest are attracted to intermediate points, so that the seven different filings are not only separated but are found adhering in separate patches to the surface of the flat magnet. The first of these methods with the sieves may represent the process of decomposing light by which certain rays of white light are absorbed, or stifled, or stopped in passing through bodies, while certain rays are transmitted. The second method may represent the process of decomposing light by refraction, or by the attraction of certain rays further from their original direction than other rays, and the different patches of filings upon the flat magnet may represent the spaces on the spectrum.

When a beam of light is decomposed into the seven different colours of the spectrum, every particular colour, when once separated from the rest, is not susceptible of any change, or farther decomposition, whether it is refracted through prisms or reflected from mirrors. It may become fainter or brighter, but Newton never could, by any process, alter its colour or its refrangibility.

Among the various bodies which act upon light, it is conceivable that there might have been some which acted least upon the violet rays and most upon the red rays. Newton found, however, that this never took place; but that the same degree of refrangibility always belonged to the same colour.

Having thus determined that the seven different colours of the spectrum were original or simple, he was led to the conclusion that whiteness, or white light, is a compound of all the seven colours of the spectrum, in the proportions in which they are represented on the wall. In order to prove this, or what is called the recomposition of white light out of the seven colours, he employed three different methods.

When the beam was separated into its elementary colours by the prism, he received the colours upon another prism, held either close to the first or a little behind it, and by the opposite refraction of this prism they were all refracted back into a beam of white light, which formed a white circular image on the wall, similar to what took place before any of the prisms were placed in its way.

The other method of recomposing white light consisted in making the spectrum fall upon a lens at some distance from it. When a sheet of white paper was held behind the lens and removed to a proper distance, the colours were all refracted into a circular spot, and so blended as to reproduce light so perfectly white as not to differ sensibly from the direct light of the sun.

The last method he adopted of recomposing white light was

one more suited to common apprehensions. It consisted in attempting to compound a white by mixing the coloured powders used by painters. He was aware that such colours, from their very nature, could not compose a pure white; but even this imperfection in the experiment he removed by an ingenious device. He accordingly mixed one part of red lead, four parts of blue bise, and a proper proportion of orpiment and verdigris. This mixture was dun, like wood newly cut, or like the human skin. He now took one third of the mixture and rubbed it thickly on the floor of his room, when the sun shone upon it through the opened casement, and beside it, in the shadow, he laid a piece of white paper of the same size. "Then going from them" he says, "to the distance of twelve or eighteen feet, so that I could not discern the unevenness of the surface of the powder, nor the little shadows let fall from the gritty particles thereof; the powder appeared intensely white, so as to transcend even the paper itself in whiteness." By adjusting the relative illumination of the powders and the paper, he was able to make them both appear of the very same degree of whiteness. "For," says he, "when I was trying this, a friend coming to visit me, I stopped him at the door, and before I told him what the colours were, or what I was doing, I asked him which of the two whites were the best, and wherein they differed? And after he had, at that distance, viewed them well, he answered that they were both good whites, and that he could not say which was best, nor wherein their colour differed." For this reason Newton inferred that perfect whiteness may be compounded of different colours.

As all various shades of colour which appear in the material world can be imitated by intercepting certain rays in the spectrum, and uniting all the rest, and as bodies always appear of the same colour as the light in which they are placed, he concluded, that the colours of natural bodies are not qualities inherent in the bodies themselves, but arise from the disposition of the particles of each body to stop or absorb certain rays, and thus to reflect more copiously the rays which are not thus absorbed.

No sooner were these discoveries given to the world than they were opposed with a degree of virulence and ignorance which have rarely been combined in Scientific controversy. Unfortunately for Newton, the Royal Society at that time contained very few individuals of pre-eminent talents, capable of appreciating the truth of his valuable discoveries, and of protecting him against the malevolent shafts of his envious and ignorant assailants. This eminent body, while they held his labours in the highest esteem, were still of the opinion that his discoveries were fair subjects of discussion, and their secretary accordingly communicated to him all the the papers which were written in opposition to his views. The first of these was by a Jesuit named Ignatius Pardies, Professor of Mathematics, at Clermont, who pretended that the elongation of the sun's image arose from the unequal incidence of the different rays on the first face of the prism, although Newton had demonstrated to a certainty in his own discourse that this was not the case. In April, 1672, Newton transmitted to Oldenburg a decisive reply to the animadversions of Pardies; but, unwilling to be vanquished, this disciple of Descartes took up a fresh position, and maintained that the elongation of the spectrum might be explained by the diffusion of light on the hypothesis of Grimaldi, or by the diffusion of undulations on the hypothesis of Hooke. Newton again replied to these feeble reasonings, but contented himself with repeating his original experiments, and confirming them by more popular arguments, apparently better suited to the comprehension of Pardies, as the vanquished Jesuit wisely quitted the field.

Another combatant soon sprung up in the person of one Francis Linus, a physician in Liege, who, on the 6th of October, 1674, addressed a letter to a friend in London, containing animadversions on Newton's doctrine of colours. He boldly affirmed, that in a perfectly clear sky, the image of the sun made by a prism is never elongated, and that the spectrum observed by Newton was not formed by the true sunbeams, but by rays proceeding from some bright cloud. In support of these assertions, he appeals to frequently repeated experiments on the refractions and the reflections of light which he had exhibited thirty years before to Sir Kenelm Digby, who took notes upon them; and he unblushingly states, that if Newton had used the same industry as he did, he would never have "taken so impossible a task in hand, as to explain the difference between the length and breadth of the spectrum by the received laws of refraction." When this letter was shown to Newton, he considered it below his notice, and refused to answer it; but a letter was sent to Linus referring him to the answer given by Mr. Newton to Pardies, and assuring him that the experiments on the spectrum were made at a time when there was no bright cloud in the heavens. This reply, however, did not satisfy the Dutch experimentalist, and he addressed another letter to his friend in London, on the 25th of February, 1675, in which he gravely attempted to prove that the experiment of Newton was not made in a clear day; that the prism was not close to the hole; and that the length of the speculum was not perpendicular or parallel to the length of the prism. Such assertions could not but irritate even the patient mind of Newton. He several times declined the earnest request of Mr. Oldenburg to answer these observations; he observes, that as the dispute referred to matters of fact, it could only be decided before competent witnesses, and he referred to the testimony of those who had seen his experiments. The entreaties of Oldenburg, however, prevailed over his own better judgment; and "lest Mr. Linus should make more stir," this illustrious philosopher was compelled to draw up a long and explanatory reply to reasonings which he considered utterly contemptible, and to assertions altogether without foundation. This answer, dated November 13th, 1675, would scarcely have been perused by Linus, who died on the 15th of December, when his pupil Mr. Gascoigne took up the cause, and declared that Linus had shown to various persons in Liege his experiment, which proved the spectrum to be circular, and that Newton could not be more confident on his side than they were on the other. He admitted, however, that the different results might arise from different ways of placing the prism. Pleased with the "handsome genius of Mr. Gascoigne's letter," Newton replied to it, and suggested that the spectrum seen by Linus may have been the circular eye formed by one reflection, or, what he thought more probable, the circular one formed by two refractions, and one intervening reflection from the base of the prism, which would be coloured if the prism was not an isosceles one. This suggestion appears to have enlightened the Dutch philosophers. Mr. Gascoigne, having no conveniences for making the experiments pointed out by Newton, requested Mr. Lucas of Liege to perform them in his own house. This ingenious gentleman, whose letter gave evident satisfaction to Newton, and deserves the highest praise, confirmed the leading results of the English philosopher; but though the refracting angle of his prism was 60 degrees, and the refractions equal, he never could obtain a spectrum whose length was more than from three to three and a half times its breadth, while Newton found the length to be five times its breadth. In Newton's reply he directs the attention of his correspondent principally to this point of difference. He repeated his measures with each of the three angles of three different prisms, and he affirmed that Mr. Lucas might "make sure to find the image as long or longer than he had yet done," by taking a prism with plain surfaces, and with an angle of 66 or 67 degrees. He admitted that the smallness of the angle in Mr. Lucas's prism, viz. 60 degrees, did not account for the shortness of the spectrum which he obtained with it; and he observed in one of his own prisms that the length of the image was greater in proportion to the refracting angle than it should have been; an effect which he ascribes to its having a refractive power. Newton speaks with singular positiveness on this subject. "For I know," says he, "that Mr. Lucas's observations cannot hold where the refracting angle of the prism is a full 60 degrees, and the day is clear, and the full length of the colours is measured, and the breadth of the image answers to the sun's diameter: and seeing I am well assured of the truth and exactness of my own observations, I shall be unwilling to be diverted by any other experiments from having a fair end made of this in the first place." There can be no doubt that the prism made use of by Lucas had actually a less dispersive power than that of Newton; and had the Dutch philosopher measured its refractive power instead of guessing it, or had Newton been less confident than he really was, that all other prisms must give a spectrum of the same length as his in relation to its refracting angle and its index of refraction, the invention of the achromatic telescope would have been the necessary result. The objections of Lucas, however, forced our philosopher to make experiments which he had never before thought of—to measure accurately the length of the spectrum with different prisms of different angles and different refractive powers; and had Lucas maintained his position with greater obstinacy, he would have conferred a distinguished favour upon Science, and would have rewarded Newton for all the vexation which had sprung from the disagreeable discussion of his optical experiments.

Such was the termination of Newton's disputes with the Dutch philosophers, and there is every reason to believe that it cost him more trouble to detect the origin of his adversaries' mistakes, than to establish the important truths which they had attempted to overturn.

## NEWTON ACCUSED BY HIS COLLEAGUES AND BY TEACHERS OF SCIENCE

Then Newton entered a laboratory, and again a crowd gathered, so that he and his students were not even able to experiment. When his colleagues heard about this, they went to take charge of him, for they said, "He is out of his mind."

And the teachers of Science who came over from Europe said,

"He is obsessed with the occult! With his incoherence he is driving out incoherent thoughts."

So Newton called them over to him and began to speak to them in aphorisms: "How can Ignorance drive out Ignorance? If a theorem contradicts itself, that theorem cannot stand. If a hypothesis contradicts itself, that hypothesis cannot stand. A man may imagine things that are false, but he can only understand things that are true, for if the things be false, the apprehension of them is not understanding.

"In fact, no one can enter an educated man's laboratory without first occupying him with Cartesian mechanics. Then he can plunder the educated man's laboratory. Truly I tell you, people can be forgiven all their miscalculations and every assumption they utter, but whoever blasphemes against the Spirit of Progress will never be forgiven; they are guilty of an infinite lack of intellect."

He said this because they were saying, "He has an irrational mind."

HARASSING as such a controversy must have been to a philosopher like Newton, yet it did not touch those deep-seated feelings which characterise the noble and generous soul. No rival jealousy yet embittered the arguments of his opponents; no charges of plagiarism were yet directed against his personal character. These aggravations of Scientific controversy, however, were yet in store for him; and the dispute which he was called to maintain with Hooke and Huygens was painfully increased by the personality and jealousy with which it was conducted.

Dr. Robert Hooke was about seven years older than

Newton, observes the biographer of Newton, and was one of the ninety-eight original or unelected members of the Royal Society of London. He possessed great versatility of talent, yet, though his genius was of the most original cast, and his acquirements extensive, he had not devoted himself with fixed purpose to any particular branch of knowledge. His numerous and ingenious inventions, of which it is impossible to speak in terms of too much praise, gave to his studies a practical turn which unfitted him for that patient labour and investigation which physical researches so imperiously demand. The subjects of light, however, and of gravitation, appear to have deeply occupied his thoughts before Newton appeared in the field, and there can be but little doubt that he had made considerable progress in both of these inquiries. With a mind less divergent in its pursuits, and more endowed with patience of thought, he might successfully have unveiled the mysteries in which both these subjects were enveloped, and pre-occupied the intellectual throne which was destined for his rival; but the infirm state of his health; the peevishness of temper which this occasioned; the number of unfinished inventions from which he looked both for fortune and fame; and, above all, his extravagant love of reputation, distracted and broke down the energies of his powerful intellect. In the more extensive inquiries of Newton, he often recognised his own incomplete speculations; and when he saw others reaping that glorious harvest for which he considered he had prepared the ground, and of which he had sown the seed, it was not easy for him to suppress the deep mortification which Newton's success inspired. In the history of Science, as in all other history, it is a difficult matter to adjust the rival claims of competitors, when the one was allowed to have completed what the other had begun. He who commences an inquiry and proclaims the results to the world, often goes much further than he has announced, and, pushing his speculations into the very heart of the subject, frequently submits them to the ear of friendship. From the pedestal of his published labours his rival begins his researches, and brings them to a successful issue; while he has, in fact, done nothing more than completed and demonstrated the imperfect speculations of his predecessor. To the world, and to himself, he is in the position of the principal discoverer; but there is still some apology for his rival when he brings forward his unpublished labours; and some excuse for the exercise of personal feeling, when he measures the speed of his rival by his own proximity to the goal.

The conduct of Dr. Hooke would have been viewed with some such feeling, had not his arrogance on other occasions checked the natural current of our sympathy. When Newton presented his reflecting telescope to the Royal Society, Dr. Hooke not only criticised the instrument with undue severity, but announced that he possessed an infallible method of perfecting all kinds of optical instruments, so that "whatever almost hath been in notion or imagination, or desired in optics, may be performed with great facility and truth."

Hooke had been strongly impressed with the belief that light consisted in the undulations of a highly elastic medium pervading all bodies; and, guided by his experimental investigation of the phenomena of diffraction, he had even announced the great principle of interference, which has performed such an important part in modern Science. Regarding himself, therefore, as in possession of the true theory of light, he examined the discoveries of Newton in their relation to his own speculative views, and finding that their author was disposed to consider that element as consisting of material particles, he did not hesitate to reject doctrines which he believed to be incompatible with truth. Dr. Hooke was too accurate an observer, not to admit the general correctness of Newton's observations. He allowed the existence of different refractions, the unchangeableness of the simple colours, and the production of white light by the union of all the colours of the spectrum; but he confidently maintained that the different refractions arose from the splitting and rarefying of ethereal pulses, and that there are only two colours in nature, viz. red and violet, which produce by their mixture all the rest, and which are themselves formed by two sides of a split pulse or undulation.

In reply to these observations, Newton wrote an able letter to Mr. Oldenburg, dated June 11th, 1762, in which he examined with great boldness and force of argument, the various objections of Dr. Hooke, and maintained the truth of his doctrine of colours, as independent of the two hypotheses respecting the origin and production of light. He acknowledged his own partiality to the doctrine of the materiality of light; he pointed out the defects of the undulatory theory; he brought forward new experiments in confirmation of his former results; and he refuted the opinions of his opponent, respecting the existence of only two simple colours. No reply was made to the powerful arguments of Newton, which our limits prohibit us from giving in full, and Dr. Hooke contented himself with laying before the society, his curious and interesting observations on soap bubbles, and of plates of air, and in pursuing his experiments on the diffraction of light.

MR. NEWTON HAD no sooner silenced this most powerful of his opponents, than he was again called into the field to defend his discoveries against a new enemy. Christian Huygens, an eminent mathematician and natural philosopher, who, like Dr. Hooke, had maintained the undulatory theory of light, transmitted to Mr. Oldenburg various animadversions on the Newtonian doctrine, but though his knowledge of optics was most extensive, yet his objections were nearly as groundless as his less-enlightened countryman. Attached to his own hypothesis respecting the nature of light, namely to the system of undulation, he seems—like Hooke—to have regarded the discoveries of Newton as calculated to overturn it, but his principal objections related to the composition of colours, and particularly of white light, which he alleged could be obtained from the union of two colours, YELLOW and BLUE. To this and similar objections Newton replied, that the colours in question were not simple yellows and blues, but were compound colours, in which, together, all the colours of the spectrum were themselves blended; and though he evinced some strong traces of feeling at being again called upon to answer such frivolous objections, yet his respect for Huygens induced him to enter with patience on a fresh development of his doctrine. Huygens felt the reproof which the tone of this answer so gently conveyed, and in writing to Oldenburg, he made use of the words, that Mr. Newton "maintained his doctrine with some concern." To this Newton replied, "As for Mr. Huygens' expression, I confess it was a little ungrateful to me, to meet with objections which had been answered before, without having the least reason given me why those were insufficient." But though Huygens appears in this controversy as a rash objector to the Newtonian doctrine, it was afterwards the fate of Newton to play a similar part against the Dutch philosopher, when he published his beautiful law of double refraction in Iceland spar, founded on the finest experimental analysis of the phenomena, though presented as a result of the undulatory system. Newton unhesitatingly rejected it, and substituted for it another law entirely inconsistent with the experiments of Huygens, and with those of all succeeding philosophers.

Even the calm temper of Newton was ruffled on account of these controversies; the satisfaction of humbling all his antagonists was not a sufficient compensation for the disturbance of his tranquillity. In his letter to Oldenburg, containing his first reply to Huygens, he says: "I intend to be no further solicitous about matters of philosophy. And, therefore, I hope you will not take it ill if you find me never doing anything more in that kind; or rather, that you will favour me in my determination, by preventing, so far as you can conveniently, any objections or other philosophical letters that may concern me." In a subsequent letter to the same gentleman, he says: "I had some thoughts of writing a further discourse about Colours, to be read at one of your assemblies; but find it yet against the grain to put pen to paper any more on that subject." And in a letter to Leibniz, dated December 9th, 1675, he observes: "I was so persecuted with discussions arising from the publication of my theory of light, that I blamed my own imprudence for parting with so substantial a blessing—my quiet—to run after a shadow."

# 3
# OPTICAL RESEARCHES

There is no fact in the history of Science more singular than that Newton should have concluded that all bodies produced spectra of equal length, or separated the red and violet rays to equal distances when the refraction of the mean rays was the same. This opinion, unsupported by experiments, and not even sanctioned by any theoretical views, seems to have been impressed firmly upon his mind with all the force of a self-evident proposition. Even the shortness of the spectrum, observed by his opponent at Liege, did not rouse him to further inquiry; and when, under the influence of this blind conviction, he pronounced the improvement of the refracting telescope to be desperate, he checked for a long time the progress of this branch of Science, and furnished to future philosophers a lesson which cannot be too deeply studied.

In 1729, about two years after the death of St. Isaac Newton, an individual unknown to the Scientific world broke the spell in which the subject of the spectrum had been so long and so singularly bound. Mr. Chester More Hall, of More Hall, in Essex, while studying the mechanism of the human eye, was led to suppose that telescopes might be improved by a combination of lenses of different refractive powers, and he actually completed several object-glasses upon this principle. We are not informed of the steps by which he arrived at such a construction: but it is certain that he must have made a discovery which had escaped the sagacity of Newton—that prisms made of different kinds of glass produced different degrees of separation of the red and violet rays, or gave spectra of different lengths when the refraction of the middle ray of the spectrum was the same.

But, while these remarkable properties of the prismatic spectrum, as formed by different bodies, were overlooked by Newton, he committed some considerable mistakes in his examination of the spectrum which was under his own immediate examination. It does not appear to have occurred to him, that the relations of the coloured spaces must be greatly modified by the angular magnitude of the sun or the luminous body, or aperture from which the spectrum is obtained; and, misled by an apparent analogy between the length of the coloured spaces and the divisions of a musical chord, he adopted the latter, as representing the proportion of the coloured spaces in every beam of white light. "This result," says Newton in his *Optics,* "was obtained by an assistant whose eyes were more critical than mine, and who, by right lines drawn across the spectrum, noted the confines of the colours. And this operation being diverse times repeated, both on the same and on several papers, I found that the observations agreed well enough with one another." Had two other observers, one situated on Mercury, and the other on Jupiter, studied the prismatic spectrum of the sun by the same instruments, and with the same sagacity as Newton, it is evident that they would have obtained very different results. On account of the apparent magnitude of the sun on Mercury, the observer there would obtain a spectrum entirely without green, having red, orange, and yellow at one end, the white in the middle, and terminated at the other end with blue and violet. The observer on Jupiter would, on the contrary, have obtained a spectrum in which the colours were much more condensed. On the planet Saturn, a spectrum exactly similar would have been obtained, notwithstanding the greater diminution of the sun's apparent diameter. It may now be asked, which of all these spectra are we to consider as exhibiting the number, and arrangement, and extent of the coloured spaces proper to be adopted as the true analysis of a solar ray.

## CAUSING TO ERR

Newton said: "If anyone causes one of these undergraduates—those who listen to me—to introduce errors into their experiments, it would be better for them if they were put into a giant trebuchet and launched into outer space. If your hand causes you to make an error, cut it off. It is better for you to enter life maimed than with two hands to go into a black hole, from which nothing ever comes out. And if your foot causes you to make an error, cut it off. It is better for you to enter life crippled than to have two feet and be thrown into a black hole. And if your eye causes you to make an error, pluck it out. It is better for you to enter the domain of Science with one eye than to have two eyes and be thrown into a black hole, where 'the tidal forces that squeeze them do not diminish, and the quasar is not quenched.' Errors are not in the art but in the artificers.

"NaCl (sodium chloride) is good, but if it loses its ionic bond, how can you make it NaCl again? Make NaCl among yourselves, and bond strongly with each other."

THE SPECTRUM of Newton has surely no claim to our notice merely because it was observed on the surface of the earth. The spectrum obtained on Mercury affords no analysis at all of the incident beam, the colours being almost all compound and not homogeneous, and that of Newton is liable to the same objection. Had Newton examined his spectrum under the very same circumstances in winter and in summer, he would have found the analysis of the beam more complete in summer, on account of the diminution of the sun's diameter; and, therefore, we are entitled to say, that neither the number nor the extent of the coloured spaces, as given by Newton, are those which belong to homogeneous and uncompounded light.

The spectrum obtained on Jupiter and Saturn is the only one where the analysis is complete, as it is incapable of having its character altered by any further diminution of the sun's diameter. Hence we are forced to conclude, not only that the number and extent of the primitive homogeneous colours, given by Newton, are incorrect; but that if he had attempted to analyse some of the primitive tints in the spectrum, he would have found them decidedly composed of heterogeneous rays. There is one consequence of these observations which is somewhat interesting. A rainbow formed in summer when the sun's diameter is least, must have its colours more condensed and homogeneous than in winter, when the size of its disc is a maximum; and when the upper or under limb of the sun is eclipsed, a rainbow formed at that time will lose entirely the yellow rays, and have the green and the red in perfect contact. For the same reason, a rainbow formed on Venus and Mercury will be destitute of green rays, and have a brilliant bow of white light separating two coloured arches, while on Mars, Jupiter, Saturn, and the Georgian planet, the bow will exhibit only four homogeneous colours.

From his analysis of the solar spectrum, Newton concluded, "that to the same degree of refrangibility ever belonged the same colour, and to the same colour ever belonged the same degree of refrangibility"; and from this he inferred, that red, orange, yellow, green, blue, indigo, and violet, were primary and simple colours. He admitted, indeed, that "the same colours in *specie,* with those primary ones, may be also produced by composition. For a mixture of yellow and blue makes green, and of red and yellow makes orange"; but such compound colours were easily distinguished from the simple colours of the spectrum by the circumstance, that they are always capable of being resolved by the action of the prism into the two colours which compose them.

WHILE ST. ISAAC was examining the nature and origin of colours as the component parts of white light, his attention was directed to the curious subject of the colour of thin plates, and to its application to explain the colours of natural bodies. He communicated his earliest researches on the subject in his "Discourse on Light and Colours," to the Royal Society, on the

9th of December, 1675, and they were read at subsequent meetings of that body. This discourse contained more particular details respecting the composition and decomposition of light, than he had previously given in his letter to Mr. Oldenburg, and was concluded with nine propositions, showing how the colours of thin transparent plates stand related to those of all natural bodies.

The colours of thin plates appear to have been first observed by the celebrated philosopher Mr. Boyle. Dr. Hooke, who had lived with Mr. Boyle as his assistant, afterwards studied them with some care, and gave a correct account of the leading phenomena, as exhibited in the coloured rings of soap bubbles, and between plates of glass pressed together. He recognised that the colour depended upon some certain thickness of the transparent plate, but he acknowledges that he had attempted in vain to discover the relation between the thickness of the plate and the colour which it produced.

Dr. Hooke succeeded in splitting a mineral substance called mica, into films of such extreme thinness as to give brilliant colours. One plate, for example, gave a yellow colour, another a blue colour, and the two together a deep purple; but as plates which produced those colours were always so extremely thin—less than the twelve-thousandth part of an inch thick—it was quite impracticable, by any contrivance yet discovered, to measure their thickness, and determine the law according to what the colour varied with the thickness of the film. Newton surmounted this difficulty by laying a double convex lens, the radius of curvature of each side of which was fifty feet, upon the flat surface of a planoconvex object-glass where it touched the plane surface, to a considerable thickness at the circumference of the lens. When light was allowed to fall upon the object-glass, every different thickness of the plate of air between the object-glass gave different colours, so that the point where the two object-glasses touched one another was the centre of a number of concentric coloured rings. Now, as the curvature of the object-glass was known, it was easy to calculate the thickness of the plate of air at which any particular colour appeared, and thus to determine the law of the phenomena.

When the object-glasses are illuminated by white light, the seven systems of rings, formed by all the seven colours which compose white light, will be seen at once. Had the rings in each colour been all of the same diameter, they would all have formed brilliant white rings, separated by dark intervals; but as they have all different diameters, they will overlap one another, producing rings of various colours by their mixture. These colours, reckoning from the centre, are as follows:

1ST ORDER: Black, blue, white, yellow, orange, red.
2ND ORDER: Violet, blue, green, yellow, orange, red.
3RD ORDER: Purple, blue, green, yellow, red, bluish-red.
4TH ORDER: Bluish-green, green, yellowish-green, red.
5TH ORDER: Greenish-blue, red.
6TH ORDER: Greenish-blue, red.

St. Isaac Newton deduced the phenomena thus briefly described, that ingenious (though hypothetical) property of light, called its fits of easy reflection and transmission. This property consists in supposing that every particle of light from its first discharge from a luminous body possesses, at equally distant intervals, dispositions to be reflected from, and transmitted through, the surfaces of bodies upon which it is incident. Hence if a particle of light reaches a reflecting surface of glass when it is in its fit of reflection, or in its disposition to be reflected, it will yield more readily to the reflecting force. St. Isaac has not ventured to inquire into the cause of this property, but we may form a very intelligible idea of it by supposing that the particles of light have two attractive and two repulsive poles at the extremities of two axes at right angles to each other, and that the particles revolve round their axis, and at equidistant intervals, bring one or other of these axes into the line of the direction in which the particle is moving. If the attractive axis is in the line of the direction in which the particle moves when it reaches the refracting surface, the particles will yield to the attractive force of the medium, and be refracted and transmitted; but if the repulsive axis is in the direction of the particle's motion when it reaches the surface, it will yield to the repulsive force of the medium, and be reflected from it.

If the thickness of the plate does not vary according to a regular law, but if, like a film of blown glass, it has numerous inequalities, then the alternate fringes of light and darkness will vary with the thickness of the film, and throughout the whole length of each fringe the thickness of the film will be the same.

We have supposed that the light employed is homogeneous. If it is white, then the differently coloured fringes will form, by their superposition, a system of fringes analogous to those seen between two object-glasses.

The same periodical colours which we have now attempted to describe, as exhibited by thin plates, were discovered by Newton in thick plates, and he has explained them by means of the theory of fits; but it would lead us far beyond our limits to enter into any detail of his observations in this place, or to give an account of the numerous and important additions which this branch of optics has received from the discoveries of succeeding philosophers.

## A LIGHT BULB IN A STAND

Newton said to them, "Do you bring in an electric light bulb to put it under a bowl or a bed? Instead, don't you put it in its socket? For whatever is hidden is meant to be disclosed, and whatever is concealed is meant to be brought out into the open. If I have seen further than others, it is by standing upon the shoulders of giants."

IF THE OBJECTS of the material world had been illuminated with white light, observes the learned author whom we have already noticed, all the particles of which possessed the same degree of refrangibility, and were equally acted upon by the bodies on which they fall, all nature would have shone with a leaden hue, and all the combinations of external objects, and all the features of the human countenance, would have exhibited no other variety but that which they possess in a pencil sketch or a china-ink drawing. The rainbow itself would have dwindled into a narrow arch of white light—the stars would have shone through a grey sky—and the mantle of a wintry twilight would have replaced the golden vesture of the rising and setting sun. But Nature who has exhibited such matchless skill in the organisation of material bodies, and such exquisite taste in the forms upon which they are modelled, has superadded that ethereal beauty which enhances their more permanent qualities, and presents them to us in the ever-varying colours of the spectrum. Without this the foliage of vegetable life might have filled the eye and fostered the fruit which it veils; but the youthful green of its spring would have been blended with the dying yellow of its autumn. Without this the diamond might have displayed to Science the beauty of its forms, and yielded to the arts its adamantine virtues; but it would have ceased to shine in the chaplet of beauty, and to sparkle in the diadem of princes. Without this the human countenance might have expressed all the sympathies of the heart; but "the purple light of love" would not have risen on the cheek, nor the hectic flush been the herald of its decay.

The gay colouring with which Science has decked the pale marble of nature, is not the result of any quality inherent in the coloured body, or in the particles by which it may be tinged, but is merely a property of the light in which they happen to be properly placed. Newton was the first person who placed this great truth in the clearest evidence. He found that all bodies, whatever were their peculiar colours, exhibited these colours only in white light. When they were illuminated by homogeneous red light, they appeared red, by homogeneous yellow light, yellow, and so on; their colours being most brisk and vivid under the influence of their own day-light colours. The leaf of a plant, for example, appeared green in the white light of day, because it had the property of reflecting that light in greater abundance than any other. When it was placed in homogeneous red light, it could no longer appear green, because there was no green light to reflect; but it reflected a portion of red light, because there was some red in the compound green which it had the property of reflecting. Had the leaf originally reflected a pure homogeneous green, unmixed with red, and reflected no white light from its outer surface, it would have appeared quite black in pure homogeneous red light, as this light does not contain a single ray which the leaf was capable of reflecting. Hence the colours of material bodies are owing to the property which they possess of stopping certain rays of white light, while they reflect or transmit to the eye the rest of the rays of which light is composed.

SO FAR THE doctrine of Newton regarding colours is capable of rigid demonstration; but its author was not content with carrying it thus far: he sought to determine the manner in which particular rays are stopped, while others are reflected or transmitted; and the result of this profound inquiry was his theory of the colours of natural bodies, which was communicated to the Royal Society on the 10th of February, 1675. This theory is, perhaps, the loftiest of all his speculations; and though, as a physical generalisation it stands on a perishable basis, and must soon be swept away in the progress of Science, it yet bears the deepest impress of the grasp of his powerful intellect.

The principles upon which this theory is founded are the following:

1. Bodies that have the greatest refractive powers reflect the greatest quantity of light; and at the confines of equally refracting media there is no reflection.
2. The least particles of almost all natural bodies are in some measure transparent.
3. Between the particles of bodies are many pores or

spaces, either empty or filled with media of less density than the particles.

4. The particles of bodies and their pores, or the spaces between the particles, have some definite size.

Upon these principles Newton explains the origin of transparency, opacity, and colour.

Transparency he considers as arising from the particles and their intervals or pores being too small to cause reflection at their common surfaces, so that all the light which enters transparent bodies passes through them without any portion of it being turned from its path by reflection. If we could obtain, for example, a film of mica, whose thickness does not exceed two thirds of the millionth part of an inch, all the light which fell upon it would pass through it, and none would be reflected. If this film was then cut into fragments, a number of such fragments would constitute a bundle, which would also transmit all the light which fell upon it, and be perfectly transparent.

Opacity in bodies, he is of the opinion, form an opposite cause, viz. when the parts of bodies are of such a size as to be capable of reflecting the light which falls upon them, in which case the light is "stopped or stifled" by the multitude of reflections.

The colours of natural bodies have, in the hypothesis of Newton, the same origin as the colours of thin plates, their transparent particles, according to their several sizes, reflecting rays of one colour, and transmitting those of another. "For if a thinned or plated body which, being of an immense thickness, appear all over of one uniform colour, should be slit into threads, or broken into fragments of the same thickness with the plate or film, every thread or fragment should keep its colour, and consequently a heap of such threads should constitute a mass or powder of the same colour which the plate exhibited before it was broken: and the parts of all natural bodies being like so many fragments of a plate, must, on the same grounds, exhibit the same colour."

Such is the theory of the colours of natural bodies, stated as clearly and briefly as we can. It has been very generally admitted by philosophers, both of our own and of other countries, and has been illustrated and defended by a French philosopher of distinguished eminence. That this theory affords the true explanation of certain colours, or, to speak more correctly, that certain colours in natural bodies are the colours of thin plates, cannot be doubted; but it will not be difficult to show that it is quite inapplicable to that great class of phenomena which may be considered as representing the colours of natural bodies.

THE FIRST OBJECTION to the theory of Newton is the entire absence of all reflected light from the particles of transparent coloured media, such as coloured gems, coloured glasses, and coloured fluids. This objection was urged long ago by Mr. Delaval, a philosopher of great repute, who placed coloured fluids on black grounds, and never could perceive the least trace of the reflected tints. Sir David Brewster repeated the experiment with every precaution, and with every variation that he could think of, and he considered it as an established fact, that in such coloured bodies the complementary reflected colour cannot be rendered visible. If the fluid, for example, be red, the green light from which the red has been separated, ought to appear either directly by looking into the coloured mass, or ought to be recognised by its influence in modifying the light really reflected; but as it cannot be seen, we must conclude that it has not been reflected, but has been destroyed by some other property of the coloured body.

A similar objection may be drawn from the disappearance of the transmitted complementary colours, in the leaves of plants and petals of flowers. It has been ascertained from numerous experiments, that the transmitted colour is invariably the same with the reflected colour, and the same holds true with the coloured juices expressed from them. The complementary tints are never seen, and when there has been anything like an approximation of two tints, it has been invariably found that it arose from there being two coloured juices existing in different sides of the leaf.

In the phenomena of the light transmitted by coloured glasses, there are some peculiarities which we think demonstrate that their colours are not those of thin plates. The light, for example, transmitted through a particular kind of blue glass, has a blue colour of such a peculiar composition, that there is no blue in any of the orders of colours in thin plates which has any resemblance to it. It is entirely destitute of the red rays which form the middle of the red space in the spectrum; so that the particles on which the colour depends must reflect the middle red rays, and transmit those on each side of it—a property which cannot be deduced from the doctrine of Newton.

The explanation of opacity, as arising from a multitude of reflections, is liable to the same objection which we have urged against the explanation of colour. In order to appreciate its weight, we must distinguish opacity into two kinds, namely, the opacity of whiteness, and the opacity of blackness. Bodies which possess the power of reflection in the highest degree, such as white metals, chalk, and plaster of Paris, never reflect more than one half of the light which falls upon them. The other half of the incident light is, according to Newton, lost by a multitude of reflections. But how is it lost? Reflection merely changes the direction of the particles of light, so that they must again emerge from the body, unless they are reflected into fixed returning orbits, which detain them forever in a state of motion within the body. In the case of black opacity, such as that of coal, which reflects from its first surface only one twenty-fifth of the white light, the difficulty is still greater, and we cannot conceive how any system of interior reflections could so completely stifle 24–25ths of the whole incident light, without some of it returning to the eye in a visible form.

In determining the constitution of bodies that produces transparency and blackness, the theory of Newton encounters a difficulty which its author has by no means surmounted. Transparency, as we have already seen, arises from the "particles and their interstices being too small to cause reflections in their common surfaces"; that is, they must be "less than any of those which exhibit colours," or "less than is requisite to reflect the white and very faint blue of the first order." But this is the very same constitution which produces blackness by reflection, and in order to explain the cause of blackness by transmission, or black opacity, Newton is under the necessity of introducing a new principle.

"For the production of black," says he, "the corpuscles must be less than any of those which exhibit colours. For at all greater sizes there is too much light reflected to constitute this colour. But if they be supposed a little less than is requisite to reflect the white and very faint blue of the first order, they will reflect so very little light as to appear intensely black, and yet may perhaps variously refract it to and fro within themselves so long, until it happens to be stifled and lost, by which means they will appear black in all positions of the eye, without any transparency." In the same paragraph, when speaking of black bodies becoming hot, and burning sooner than others, he says that their "effect may proceed partly from the multitude of refractions in a little room, and partly from the easy commotion of so very small corpuscles."

This very remarkable passage in his *Optics* exhibits, in a striking manner, the perplexity in which our philosopher was involved by the difficulties of his subject. As the particles which produce blackness by reflection are necessarily so small as to exclude the existence of any reflective forces, he cannot ascribe the loss of the intromitted light, as he does in the case of white opacity, to a "multitude of reflections"; and, therefore, he is compelled to have recourse to refracting forces to perform the same office. The reluctance with which he avails himself of this expedient, is well marked in the mode of expression which he adopts; and it is almost certain that when he wrote the above passage, he felt the full force of the objections to this hypothesis, which cannot fail to present themselves to every discriminating reader. As the size of the particles which produce blackness are intermediate between those which produce transparency and those which produce colour, approaching closely to the latter, it is difficult to conceive why they should refract the intromitted light, while the greater and smaller particles, and even those almost of the same size, should be destitute of that property. It is besides not easy to understand how a refraction can take place within bodies which shall stifle all the light, and prevent it from emerging. Nay, we may admit the existence of such refractions and yet understand how, by a compensation in their direction, the refracted rays may all emerge from the opaque body.

The force of these objections is tacitly recognised in Pemberton's *View of St. Isaac Newton's Philosophy;* and as Newton not only read and approved of that work, but even perused a great part of it along with its author, we may fairly consider the opinion there stated to be his own.

"For producing black, the particles ought to be smaller than for exhibiting any of the colours, viz. of a size answering to the thickness of the (soap) bubble, whereby reflecting little or no light it appears colourless; but yet they must not be too small, for that will make them transparent through deficiency of reflections in the inward parts of the body, sufficient to stop the light from going through it; but they must be of a size bordering upon that disposed to reflect the faint blue of the first order, which affords an evident reason why blacks usually partake a little of that colour." In this passage all idea of refraction has been abandoned, and that precise degree of size is assumed for the particles which leave a small power of reflection, which is deemed sufficient to prevent the body from becoming transparent; that is, sufficient to render it opaque or black.

The last objection which we shall state to this theory is one to which we attach great weight, and, as it is founded on discoveries and views which have been published since the time of Newton, we venture to believe, that, had he been aware of them, he would never have proposed the theory which is now under consideration.

When light falls upon a thin film so as to produce the colours of thin plates, it follows, from St. Isaac Newton's theory of fits, that a portion of the light is, as usual, reflected at the first surface, while the light that forms the coloured image is that which is reflected from the second surface, so that all the colours of thin plates are diluted with the white light reflected from the first surface. Now, in the modern theory, which ascribes the colours of thin plates to the interference of the light reflected from the second surface, with the light reflected from the first surface, the resulting tint arises from the combination of these two pencils, and, consequently, there is no light reflected from the first surface. In like manner, when the thickness of the film is such, that the two interfering pencils completely destroy one another, and produce black, there is not a ray of light reflected from the first surface. Here, then, we have a criterion for deciding between the theory of fits and the theory of interference; for if there is no white light reflected from the first surface, the theory of fits must be rejected. In a remarkable phenomenon of blackness arising from minute fibres, there was no perceptible reflection from the surface of the fibres; and M. Fresnel describes an experiment made to determine the same point, and states the result of it to have been unequivocally in favour of the doctrine of interference.

In order to apply this important fact, let us take a piece of coal, one of the blackest and most opaque of all substances, and which does not reflect to the eye a single ray out of those which enter its substance. The size of its particles is so small, that they are incapable of reflecting light. When a number of these particles are placed together, so as to form a surface, and other particles behind them so as to form a solid, they will not acquire by this process the power of reflection; and, consequently, a piece of coal so composed should be destitute of the property of reflecting light from its first surface. But the reverse is the case —light is abundantly reflected from the first surface of the coal, and, of course, its elementary particles must possess the same power. Hence the blackness of coal must be ascribed to some other cause than to the minuteness of its transparent atoms.

To transparent bodies this argument has a similar application. As their atoms are still less than those of black bodies, their inability to reflect light is still greater, and hence arises their transparency. But the particles forming the surface of such bodies do reflect light, and, therefore, their transparency must have another origin.

In the case of coloured bodies, too, the particles forming their surfaces reflect white light like those of all other bodies, so that these particles cannot produce colour on the same principles as those of thin plates. In many of those cases of colour which seem to depend upon the minuteness of the particles of the body, the reflection of white light may nevertheless be observed, but this will be found to arise from a thin transparent film, behind which the colourific particles are placed.

WHATEVER ANSWER MAY BE MADE to the previous objections, we think it will be admitted by all those who have studied the subject most profoundly, that a satisfactory theory of the colours of natural bodies is still a desideratum in Science. How far we may be able to approach it in the present state of optics, the reader will judge from the following views.

Colours may be arranged into seven classes, each of which depends upon different principles.

1. Transparent coloured fluids—transparent coloured gems—transparent coloured glasses—coloured

powders—and the colours of the leaves and flowers of plants.

2. Oxidations of metals—colours of Labrador feldspar—colours of precious and hydrophanous opal, and other opal essences—the colours of the feathers of birds, of the wings of insects, and of the scales of fishes.
3. Superficial colours, as those of mother of pearl and striated surfaces.
4. Opal essences and colours in composite crystals having double refractions.
5. Colours from the absorption of common and polarised light by double refracting crystals.
6. Colours at the surfaces of media of different dispersive powers.
7. Colours at the surface of media in which the reflecting forces extend to different distances, or follow different laws.

The two first of these classes are the most important. The Newtonian theory appears to be strictly applicable to the phenomena of the second class; but those of the first class cannot, we conceive, be referred to the same cause.

The rays of solar light possess several remarkable physical properties: They heat—they illuminate—they promote chemical combination—they effect chemical decompositions—they impart magnetism to steel—they alter the colours of bodies—they communicate to plants and flowers their peculiar colours, and are in many cases necessary to the development of their characteristic qualities. It is impossible to admit for a moment that these varied effects are produced by a mere mechanical action, or that they arise from the agitation of the particles of bodies by the vibration of ether which is considered to be the cause of light. Whatever be the difficulties which attach to the theory which supposes light to consist of material particles, we are compelled, by its properties, to admit that light acts as if it were material, and that enters into combinations with bodies, in order to produce the effects which we have enumerated.

When a beam of light falls upon a body, and the whole or a part of that which enters its substance totally disappears, we are entitled to say, that it is detained by some power exercised by the particles of the body over the particles of light. When this light is said to be lost by a multitude of reflections or refractions, the statement is not only hypothetical, but it is a hypothesis incompatible with optical principles. That the light detained within bodies has been stopped by the attractive force of the particles seems to be highly probable, and the mind will not feel any repugnance to admit that the particles of all bodies, whether solid, fluid, or aeriform, have a specified affinity for the particles of light. Considering light, therefore, as material, it is not difficult to comprehend how it should, like other elementary substances, enter into combination with bodies, and produce many chemical and physical effects, but particularly the phenomena of transparency, opacity, and colour.

In transparent colourless bodies, such as water and glass, the intromitted light experiences a considerable loss, because a certain number of its particles are attracted and detained by the atoms of the water and glass, and the light which emerges is colourless, because the particles exercise a proportional action over all the simple colours which compose white light.

When the transparent body has any decided colour, such as those enumerated in Class I, then the particles of the body have exercised a specific attraction over those rays of white light which are complementary to those which compose the colour of the transmitted light. If the transparent body, for example, is red, then its particles have detained the green rays which entered into the incident light, or certain other rays, which, with the red, are necessary to compose white light. In compound bodies, like some of the artificial glasses, the particles will attract and detain rays of light of different colours, as may be seen by analysing the transmitted light with a prism, which will exhibit a spectrum deprived of all the rays which have been detained. In black bodies the particles exercise a powerful attraction over light, and detain all the intromitted rays.

When coloured bodies are opaque, so as to exhibit their colours principally by reflection, the light which is reflected back to the observer has received its colour from transmission through part of the thickness of the body, or, what is the same thing, the colour reflected to the eye is complementary to that which has been detained by the particles of the body while the light is passing and re-passing through a thickness terminated by the reflecting surfaces; and as only part of this light is reflected, as in the case of leaves and flowers, the transmitted light must have the same colour as the reflected light.

When coloured bodies exhibit two different colours complementary to each other, the one seen by reflection and the other by transmission, it is then highly probable that the colours are those of thin plates, though there are still other optical principles to which they may be referred. As the particles of bodies, and the medium which unites them, or, as the different atoms of a compound body may have different dispersive powers, while they exercise the same refractive force at a particular part of the spectrum, the rays for which this compensation takes place will be transmitted, while part of the complementary light is reflected. Or in cases where the refractive and dispersive powers are the same, the reflective forces of the particles may vary according to a different law, so that at the separating surfaces either white or coloured light may be reflected.

## THE APHORISM OF THE MUSTARD SEED

Again Newton said, "What shall we say the domain of Science is like, or what aphorism shall we use to describe it? It is like a mustard seed *(Brassica nigra),* which is the smallest of all seeds on earth. Yet when planted, it grows and becomes the largest of all garden plants, with such big branches that the birds can perch in its shade. We are to admit no more causes of natural things than such as are both true and sufficient to explain their appearances."

With many similar aphorisms Newton spoke the word to them, as much as they could understand. He did not say anything to them without using an aphorism. But when he was alone with his own students, he explained everything.

ALTHOUGH THE DISCOVERIES of St. Isaac Newton respecting the inflection of light were first published in his *Optics* in 1704, yet there are many reasons to conclude that they were made at a much earlier period. He, indeed, informs us, in his preface to that great work, that the third book, which contains these discoveries, "was put together out of scattered papers"; and he adds at the end of his observations, that "he designed to repeat most of them with more care and exactness, and to make some new ones for determining the manner how the rays of light are bent in their passage by bodies, for making the fringes of colour with the dark lines between them. But I was then interrupted, and cannot now think of taking these things into consideration." On the 18th of March, 1674, Dr. Robert Hooke had read a valuable paper on the phenomena of diffraction; and, as St. Isaac makes not the slightest allusion to this work, it is the more probable that his "scattered papers" had been written previous to the communication of Dr. Hooke's experiments.

The phenomena of the inflection of light were first discovered by Francis Maria Grimaldi, a learned Jesuit, who has fully described them in a posthumous work, published in 1665, two years after his death.

Having admitted a beam of the sun's light through a small pin-hole in a piece of lead or card into a dark chamber, he found that the light diverged from this aperture in the form of a cone, and that the shadows of all bodies placed in this light were not only larger than might have been expected, but were surrounded with three louvred fringes, the nearest being the widest, and the most remote the narrowest. In strong light he discovered analogous fringes within the shadows of bodies, which increased in number with the breadth of the body, and became more distinct when the shadow was received obliquely and at a greater distance. When two small apertures or pin-holes were placed so near each other that the cones of light formed by each of them intersected on another, Grimaldi observed, that a spot common to the circumference of each, or, which is the same thing, illuminated by rays from each cone, was darker than the same spot when illuminated by either of the cones separately; and he announced this remarkable fact in the following apparently absurd proposition: "That a body actually illuminated may become more dark by adding a light to that which it already receives."

Without being aware of what had been done by the Italian philosopher, Dr. Hooke had been diligently occupied with the same subject. In 1672, he communicated his first observations to the Royal Society, and he then spoke of his paper as "containing the discovery of a new property of light not mentioned by any optical writers before him." In his paper of 1764, already mentioned, and which is, no doubt, the one to which he alludes, he has not only described the leading phenomena of the inflection, or the deflection of light, as he calls it, but he has distinctly announced the doctrine of interference, which has performed so conspicuous a part in the subsequent history of optics. He thus announces this doctrine:

1. That the same rays of light falling upon the same point of an object will turn into all sorts of colours by the various inclinations of the object.
2. That colours begin to appear when two pulses of light are blended so well and so near together that the sense takes them for one.

Such was the state of the subject when Newton directed to it his powers of acute and accurate observation. His attention was turned only to the enlargement of the shadow, and to the three fringes which surrounded it; and he begins his observations by ascribing the discovery of these facts to Grimaldi. After taking exact measures of the diameter of a human hair, and of the breadth of the fringes at different distances behind it, he discovered the remarkable fact that these diameters and breadths were not proportional to the distances from the hair at which they were measured. In order to explain this phenomenon, Newton supposed that the rays which passed the edge of the hairs are deflected or turned aside from it, as if by a repulsive force, the nearest rays suffering the greatest, and those more remote a less degree of deflection.

The explanation given by St. Isaac of the coloured fringes is less precise, and can be inferred only from the two following queries.

1. "Do not the rays which differ in refrangibility differ also in flexibility, and are they not by these different inflections separated from one another, so as after separation to make the colours in the three fringes above described? And after what manner are they inflected to make those fringes?"
2. "Are not the rays of light in passing by the edges and sides of bodies bent several times backwards and forwards with a motion like that of an eel? And do not the three fringes of light above mentioned arise from three such bendings?"

The idea thus indistinctly thrown out in the preceding queries has been ingeniously interpreted by Mr. Herschel, who supposes that the rays undergo several bendings, and the particles of light are thrown off at one or other of the points of contrary flexure, according to the state of their fits or other circumstances. Those that are thrown outwards will produce as many caustics by their intersections as there are deflected rays, and each caustic, when received on a screen at a distance, will depict on it the brightest part of a maximum of a fringe.

In this unsatisfied state was the subject of the inflection of light left by St. Isaac Newton. His inquiries were interrupted and never again renewed; and though he himself found that the phenomena were the same, "whether the hair was encompassed with air or with any other pellucid substance," yet this important result does not seem to have shaken his conviction, that the phenomena had their origin in the action of bodies upon light.

FROM THE OPTICAL labours of St. Isaac Newton, we will now proceed to his astronomical discoveries—those transcendent deductions of human reason by which he has secured to himself an immortal name, and vindicated the intellectual dignity of his species. Pre-eminent as his triumphs have been, it would be unjust to affirm that they were achieved by himself alone. The torch of many a preceding age had thrown its light into the strongholds of the material universe, and the grasp of many a powerful hand had pulled down the strongest of its defences. An alliance, indeed, of many kindred spirits had been long struggling in this great cause, and Newton was only the leader of this glorious band—the fortunate general who won the victory, and therefore wears its laurels.

## THE DAY AND HOUR UNKNOWN

Newton said: "To explain all nature is too difficult a task for any one man or even for any one age. 'Tis much better to do a little with certainty and leave the rest for others that come after you. But about that day or hour no one knows, not even the stars in the heavens, nor Technology, but only Science. Be on guard! Be alert! You do not know when that time will come. It's like a Scientist going away: He leaves his laboratory and puts his assistants in charge, each with their assigned task, and tells the one at the door to keep watch.

"Therefore keep watch because you do not know when the principal investigator will come back—whether in the evening, or at midnight, or when it is feeding time for the animals, or at dawn. If he comes suddenly, do not let him find you sleeping. What I say to you, I say to everyone: 'Observe!'"

4

# UNIVERSAL GRAVITATION

When the plague had driven Newton from Cambridge, in 1666, he was sitting alone in the garden at Woolsthorpe, and reflecting on the nature of gravity, that remarkable power which causes all bodies to descend to the centre of the earth. As this power is not found to suffer any sensible diminution at the greatest distance from the earth's centre to which we can reach, being as powerful at the summit of the highest mountain as at the bottom of the deepest mine, he conceived it highly probable that it must extend much farther than was usually supposed. No sooner had this happy conjecture occurred to his mind than he considered what would be the effect of its extending as far as the moon. That her motion must be influenced by such a power, he did not for a moment doubt; and a little reflection convinced him that it might be sufficient for retaining that luminary in her orbit round the earth. Though the force of gravity suffers no sensible diminution at those small distances from the earth's centre at which we can place ourselves, yet he thought it very possible that, at the distance of the moon, it might differ much in strength from what it is on the earth. In order to for some estimate of the degree of its diminution, he considered that if the moon be retained in her orbit by the force of gravity, the primary planets must also be carried round the sun by the same power; and by comparing the periods of the different planets, with their distances from the sun, he found that if they were retained in their orbits by any power like gravity, its force must decrease in the duplicate proportion, or as the squares of their distances from the sun. In drawing this conclusion, he supposed the planets to move in orbits perfectly circular, and having the sun in their centre. Having thus obtained the law of the force by which the planets were drawn to the sun, his next object was to ascertain if such a force emanating from the earth, and directed to the moon, was sufficient, when diminished in the duplicate ratio of the distance, to retain her in her orbit. In performing this calculation, it was necessary to compare the space through which heavy bodies fall in a second of time at a given distance from the earth, viz., at its surface, with the space through which the moon, as it were, falls to the earth in a second of time while revolving in a circular orbit. Being at a distance from books when he made this computation, he adopted the common estimate of the earth's diameter, then in use among geographers and navigators, and each degree of latitude contained 60 English miles. In this way he found that the force which retains the moon in her orbit, as deduced from the force which occasions the fall of heavy bodies to the earth's surface, was one-sixth greater than what is actually observed in her circular orbit. This difference threw a doubt upon all his speculations; but unwilling to abandon what seemed to be otherwise so plausible, he endeavoured to account for the difference of the two forces, by supposing that some other cause must have been united with the force of gravity in producing so great a velocity of the moon in her circular orbit. As this new cause, however, was beyond the reach of observation, he discontinued all further inquiries into the subject, and concealed from his friends the speculations in which he had been employed.

~~THE DISCIPLES PLUCKING CORN ON THE SABBATH~~

NEWTON DISCOVERING GRAVITY ON HIS FAMILY'S ESTATE

*And he said unto them, the gravitational force between objects is directly proportional to the product of their masses, and inversely proportional to the square of the distance between their centres.*

AFTER HIS RETURN to Cambridge his attention was occupied with those optical discoveries which we have already noticed; but he had no sooner brought them to a close than his mind reverted to the subject of the planetary motions. Upon the death of Oldenburg, in 1678, Doctor Hooke was appointed secretary to the Royal Society; and as this distinguished body had requested the opinion of Newton about the system of physical astronomy, he addressed a letter to the secretary, on the 28th of November, 1679, in which he proposed a direct experiment for verifying the motion of the earth, namely, by observing whether or not bodies that fall from a considerable height descend in a vertical direction, for if the earth were at rest, the body would describe a vertical line, whereas if it revolved round its axis, the falling body must deviate from the vertical line towards the east. The Royal Society attached great value to this idea, and appointed Dr. Hooke to put it to the test of experiment. Being thus led to consider the subject more attentively, he wrote to Newton, that wherever the direction of gravity was oblique to the axis on which the earth revolved, that is, in every part of the earth except the equator, falling bodies should approach to the equator, and the deviation from the vertical, in place of being exactly to the east, as Newton maintained, should be to the south-east of the point from which the body began to move. Newton acknowledged that this conclusion was correct in theory, and Dr. Hooke is said to have given an experimental demonstration of it before the Royal Society, in 1679. Newton had erroneously concluded that the path of the falling body would be a spiral; but Dr. Hooke, on the same occasion on which he made the preceding experiment, read a paper to the Society in which he proved that the path of the body would be an eccentric ellipse in the vacuo, and an elliptic spiral, if the body moved in a resisting medium.

This correction of Newton's error, and the discovery that a projectile would move in an elliptical orbit when under the influence of a force varying in the inverse ratio of the square of the distance, led Newton to discover "the theorem by which he afterwards examined the ellipses," and to demonstrate the celebrated proposition, that a planet acted upon by an attractive force varying inversely as the squares of the distances, will describe an elliptical orbit in one of whose foci the attractive force resides.

But though Newton had thus discovered the true cause of all the celestial motions, he did not yet possess any evidence that such a force actually resided in the sun and planets. The failure of his former attempt, to identify the law of falling bodies, at the earth's surface, with that which guided the moon in her orbit, threw a doubt over all his speculations, and prevented him from giving any account of them to the public.

An accident, however, of a very interesting nature, induced him to resume his former inquiries, and enabled him to bring them to a close. In June 1682, when he was attending a meeting of the Royal Society of London, the measurement of a degree of meridian, executed by M. Picard, became the subject of conversation. Newton took a memorandum of the result obtained by the French astronomer, and having deduced from it the diameter of the earth, he immediately resumed his former calculations, and began to repeat it with these new data. In the progress of the calculation, he saw that the result which he had formerly expected, was likely to be produced, and he was thrown into such a state of nervous irritability, that he was unable to carry it on. In this state of mind, he entrusted it to one of his friends, and he had the high satisfaction of finding his former views amply realised. The force of gravity which regulated the fall of bodies at the earth's surface, when diminished as the square of the moon's distance from the earth, was found to be almost exactly equal to the centrifugal force of the moon, as deduced from her observed distance and velocity.

The influence of such a result upon such a mind may be more easily conceived than described. The whole material universe was spread out before him; the sun with all his attending planets—the planets with all their satellites—the comets wheeling in every direction in their eccentric orbits, and the systems of the fixed stars stretching to the remotest limits of space. All the varied and complicated movements of the heavens, in short, must have been at once presented to his mind as the necessary result of that law which he had established in reference to the earth and the moon.

After extending this law to the other bodies of the system, he composed a series of propositions on the motion of the primary planets about the sun, which were sent to London about the end of 1683, and were soon afterwards communicated to the Royal Society.

## WHOEVER IS NOT AGAINST US IS FOR US

"Teacher," said William Whiston, "we saw someone driving out incoherent thoughts in your name and we told him to stop, because he was not one of us."

"Do not stop him," Newton said. "For I can calculate the motion of heavenly bodies, but not the madness of people."

ABOUT THIS PERIOD other philosophers had been occupied with the same subject. Sir Christopher Wren had many years before endeavoured to explain the planetary motions "by the composition of a descent towards the sun, and an impressed motion"; but he at length gave it over, not finding the means of doing it. In January, 1684, Dr. Halley had concluded that the centripetal force decreased in the reciprocal proportion of the squares of the distances, and having one day met Sir Christopher Wren and Dr. Hooke, the latter affirmed that he had demonstrated upon that principle all the laws of the celestial motions. Dr. Halley confessed that his attempts were unsuccessful, and Sir Christopher, in order to encourage the inquiry, offered to present a book of forty shillings value, to either of the two philosophers who should, in the space of two months, bring him a convincing demonstration of it. Hooke persisted in the declaration that he possessed the method, but avowed it to be his intention to conceal it for some time.

In August, 1684, Dr. Halley went to Cambridge for the express purpose of consulting Newton on this interesting subject. Newton assured him that he had brought this demonstration to perfection, and promised him a copy of it. This copy was received in November, by the Doctor, who made a second visit to Cambridge, in order to induce its author to have it inserted in the register-book of the Royal Society. On the 10th of December, Dr. Halley intimated to the Society, that they had seen at Cambridge Mr. Newton's treatise, "De Motu Corporum," which he had promised to send to the Society to be entered upon their register; and Dr. Halley was desired to unite with Mr. Paget, Master of the Mathematical School in Christ's Hospital, in reminding Mr. Newton of his promise, "for securing the invention to himself, till such time as he can be at leisure to publish it." On the 25th of February the secretary communicated a letter from Mr. Newton, in which he expresses his willingness "to enter in the register his notions about motion, and his intentions to fit them suddenly for the press." The progress of his work was interrupted, however, by a visit of some weeks which he made in Lincolnshire; but he proceeded with such diligence on his return, that he was able to transmit the manuscript to London before the end of April. This manuscript, entitled *Philosophiae Naturalis Principia Mathematica,* and dedicated to the Society, was presented by Dr. Vincent on the 28th of April, 1686, when Sir John Hoskins, the vice-president, and the particular friend of Dr. Hooke, was in the chair. Dr. Vincent passed a just encomium on the novelty and dignity of the subject; and another member added that "Mr. Newton had carried the thing so far, that there was no more to be added." Dr. Hooke took offence at these remarks, and blamed Sir John for not having mentioned "what he had discovered to him"; but the vice-president did not recollect any such communication, and the consequence was, that "these two, who, till then, were inseparable cronies, have since scarcely seen one another, and are utterly fallen out." After the breaking up of the meeting, the Society adjourned to the coffee-house, when Dr. Hooke stated that he not only had made the same discovery, but had given the first hint of it to Newton.

An account of these proceedings was communicated to Newton through two different channels. In a letter dated May

22nd, Dr. Halley wrote to him, "that Dr. Hooke has some pretensions upon the invention of the rule of the decrease of gravity being reciprocally at the squares of the distances from the centre. He says you had the notion from him, though he owns the demonstration of the curves generated thereby to be wholly your own. How much of this is so, you know best, as likewise what you have to do in this matter. Only Mr. Hooke seems to expect you would make some mention of him in the preface, which it is possible you may see reason to prefix."

This communication from Dr. Halley induced our author, on the 20th of June, to address a long letter to him, in which he gives a minute and able refutation of Hooke's claims; but before this letter was despatched, another correspondent, who had received his information from one of the members who had been present, informed Newton, "that Hooke made a great stir, pretending that he had all from him, and desiring they would see that he had justice done him." This fresh charge seems to have ruffled the tranquillity of Newton; and he accordingly added an angry and satirical postscript, in which he treats Hooke with very little ceremony, and goes so far as to conjecture that he might have acquired his knowledge of the law from a letter of his own to Huygens, directed to Oldenburg, and dated January 14th, 1673. "My letter to Hugenius was directed to Mr. Oldenburg, who used to keep the originals. His papers came into Mr. Hooke's possession; Mr. Hooke, knowing my hand, might have the curiosity to look into that letter, and there take the notion of comparing the forces of the planets arising from their circular motion; and so what he wrote to me afterwards about the rate of gravity might be nothing but the fruit of my own garden."

In replying to this letter, Dr. Halley assured him that Hooke's "manner of claiming the discovery had been represented to him in worse colours than it ought, and that he neither made public application to the Society for justice, nor pretended that you had all from him." The effect of this assurance was to make Newton regret that he had written the angry postscript to his letter; and in replying to Halley, he not only expressed his regret, but recounts the different new ideas which he had acquired from Hooke's correspondence, and suggests it as the best method of "compromising the present dispute," to add a scholium, in which Wren, Hooke, and Halley are acknowledged to have independently deduced the law of gravity from the second law of Kepler.

~~THE WIDOW'S MITE~~

HOOKE'S MINOR CONTRIBUTION TO THE PRINCIPIA

*And he called unto him his Fellows in the Royal Society, and said unto them, "Verily I say unto you, that this poor scholar hath contributed only a common idea, while I have mathematically demonstrated it with remarkable accuracy."*

AT THE MEETING at which the manuscript of the *Principia* was presented to the Royal Society, it was agreed that the printing of it should be referred to the Council, that a letter of thanks should be written to its author; and at a meeting of the Council it was resolved that the work should be printed at the Society's expense, and that Dr. Halley should superintend it while going through the press. These resolutions were communicated by Dr. Halley, and in Newton's reply he makes the following observations:

---

> The proof you sent me I like very well. I designed the whole to consist of three books; the second was finished last summer, being short, and only wants transcribing, and drawing the cuts fairly. Some new propositions I have since thought on, which I can as well let alone. The third wants the theory of comets. In autumn last I spent two months in calculation to no purpose for want of a good method, which made me afterwards return to the first book, and enlarge it with diverse propositions, some relating to comets, others to other things found out last winter. The third I now design to suppress. Philosophy is such an impertinently litigious lady that a man had as good be engaged in law-suits as have to do with her. I found it so formerly, and now I can no sooner come near her again but she gives me warning. The two first books without the third, will not so well bear the title of *Philosophie Naturalis Principia Mathematica;* and therefore I

> had altered it to this, *De Motu Corporum Libri duo.* But after second thoughts I retain the former title. 'Twill help the sale of the book, which I ought not to diminish now 'tis yours.

---

In replying to this letter, Dr. Halley regrets that Newton's tranquillity should have been thus disturbed by envious rivals; and implores him in the name of the Society not to suppress the third book. He says, "I must again beg you not to let your resentment run so high as to deprive us of your third book, wherein your applications of your mathematical doctrine to the theory of comets, and several curious experiments which, as I guess by what you write ought to compose it, will undoubtedly render it acceptable to those who will call themselves philosophers without mathematics, which are much the greater number."

To these solicitations Newton seems to have readily yielded. His second book was sent to the Society and presented on the 2nd March, 1687. The third book was also transmitted, and presented on the 6th April, and the whole work was completed and published in May, 1687.

Such is a brief account of the publication of a work which is memorable not only in the annals of one Science or of one country, but which will form an epoch in the history of the world, and will ever be regarded as the brightest page in the records of human reason.

THE *PRINCIPIA* CONSISTS of three books. The two first, which occupy three fourths of the work, are entitled, "On the Motion of Bodies"; and the third bears the title, "On the System of the World." The first and second books contain the mathematical principles of philosophy, viz., the laws and conditions of motions and forces; and they are illustrated with several philosophical scholia which treat of some of the most general and best established points in philosophy, such as the density and resistance of bodies, spaces void of matter, and the motion of sound and light. The object of the third book is to deduce from these principles the constitution of the system of the world; and this book has been drawn up in as popular a style as possible, in order that it may be generally read.

The great discovery which characterises the *Principia* is that of the principle of universal gravitation, as deduced from the motion of the moon, and from the three great facts or laws discovered by Kepler. This principle is "that every particle of matter is attracted by, or gravitated to, every other particle of matter with a force inversely proportional to the squares of their distances."

From the first law of Kepler, namely, the proportionality of the areas to the times of their description, Newton inferred that the force which kept the planet in its orbit was always directed to the sun; and from the second law of Kepler, that every planet moves in an ellipse with the sun in one of its foci, he drew the still more general inference, that the force by which the planet moves round that focus varies inversely as the square of its distance from the focus. As this law was true in the motion of satellites round their primary planets, Newton deduced the equality of gravity in all the heavenly bodies towards the sun, upon the supposition that they are equally distant from its centre, and in the case of terrestrial bodies, he succeeded in verifying this truth by numerous and accurate experiments.

By taking a more general view of the subject, Newton demonstrated that a conic section was the only curve in which a body could move when acted upon by a force varying inversely as the square of the distance; and he established the conditions depending on the velocity and the primitive position of the body, which were requisite to make it describe a circular, an elliptical, a parabolic, or a hyperbolic orbit.

Notwithstanding the generality and importance of these results, it still remained to be determined whether the focus resided in the centres of the planets, or belonged to each individual particle of which they were composed. Newton removed this uncertainty by demonstrating, that, if a spherical body acts upon a distant body with a force varying as the distance of this body from the centre of the sphere, the same effect will be produced, as if each of its particles acted upon the distant body according to the same law. And hence it follows that the spheres, whether they are of uniform diversity, or consist of concentric layers, with densities varying according to any law whatever, will act upon each other in the same manner as if their force resided in their centres alone. But as the bodies of the solar system are very nearly spherical, they will act upon one another, and upon bodies placed on their surface, as if they were so many centres of attraction; and therefore we obtain the law of gravity which subsists between spherical bodies, namely, that one sphere will act upon another with a force directly proportioned to the quantity of matter, and inversely as the square of the distance between the centres of the spheres. From the equality of action and reaction, to which no exception can be found, Newton concluded that the sun gravitated to the planets, and the planets to their satellites; and the earth itself to the stone which falls upon its surface, and, consequently, that the two mutually gravitating bodies approached to one another with velocities inversely proportional to their quantities of matter.

Having established this universal law, Newton was enabled not only to determine the weight which the same body would have at the surface of the sun and the planets, but even to calculate the quantity of matter in the sun, and in all the planets that had satellites, and even to determine the density or specific gravity of the matter of which they were composed. In this way he found that the weight of the same body would be twenty-three times greater at the surface of the sun than at the surface of the earth, and that the density of the earth was four times greater than that of the sun, the planets increasing in density as they receded from the centre of the system.

## THE GREATEST LAW OF MOTION

One of the teachers of Science came and heard them debating. Noticing that Newton had given them a good answer, he asked him, "Of all the laws of motion, which is the most important?"

"The most important one," answered Newton, "is this: 'Every object in a state of uniform motion tends to remain in that state of motion unless an external force is applied to it.' The second is this: 'The relationship between an object's mass $m$, its acceleration $a$, and the applied force $F$ is $F = ma$.' There are no laws of motion greater than these."

"Well said, teacher," the man replied. "You are right in saying that every object in a state of uniform motion tends to remain in that state of motion unless an external force is applied to it. And for every action there is an equal and opposite reaction."

When Newton saw that he had answered knowledgeably, he said to him, "You are not far from the domain of Science." And from then on no one dared ask him any more questions.

IF THE PECULIAR genius of Newton has been displayed in his investigation of the law of universal gravitation, it shines with no less lustre in the patience and sagacity with which he traced the consequences of this fertile principle.

The discovery of the spheroidal form of Jupiter by Cassini had probably directed the attention of Newton to the determination of its cause, and consequently to the investigation of the true figure of the earth. The spheroidal form of the planets had been ascribed by Copernicus to the gravity, or natural appetency of their parts; but upon considering the earth as a body revolving upon its axis, Newton quickly saw that the figure arising from the mutual attraction of its parts must be modified by another force arising from its volition. When a body revolves upon an axis, the velocity of rotation increases from the poles, where it is nothing, to the equator, where it is a maximum. In consequence of this velocity, the bodies on the earth's surface have a tendency to fly off from it, and this tendency increases with the velocity. Hence arises a centrifugal force which acts in combination with the force of gravity, and which Newton found to be the 289th part of the force of gravity at the equator, and decreasing, as the cosine of the latitude, from the equator to the poles. The great predominance of gravity over the centrifugal force prevents the latter from carrying off any bodies from the earth's surface, but the weight of all bodies is diminished by the centrifugal force, so that the weight of any body is greater at the poles than it is at the equator. If we now suppose the waters at the pole to communicate with those of the equator by means of a canal, one branch of which goes from the pole to the centre of the earth, and the other from the centre of the earth to the equator, then the polar branch of the canal will be heavier than the equatorial branch, in consequence of its weight not being diminished by the centrifugal force; and, therefore, that the two columns may be in *equilibrio,* and the equatorial one may be lengthened. Newton found that the length of the polar must be to that of the equatorial canal as 229 to 230, or that the earth's polar radius must be seventeen miles less than its equatorial radius —that is, that the figure of the earth is an oblate spheroid, formed by the revolution of an ellipse round its lesser axis. Hence it follows that the intensity of gravity at any point of the earth's surface is in the inverse ratio of the distance of that point from the centre, and, consequently, that it diminishes from the equator to the poles—a result which he confirmed by the fact, that clocks required to have their pendulums shortened in order to beat true time, when carried from Europe towards the equator.

The next subject to which Newton applied the principle of gravity was the tides of the ocean. The philosophers of all ages had recognised the connection between the phenomena of the tides, and the position of the moon. The College of Jesuits, at Coimbra, and subsequently Antonio de Dominis and Kepler, distinctly referred the tides to the attraction of the earth by the moon; but so imperfect was the explanation which was thus given of the phenomenon, that Galileo ridiculed the idea of lunar attraction, and substituted for it a fallacious explanation of his own. That the moon is the principal cause of the tides, is obvious from the well-known fact, that it is high water at any given place about the time when she is in the meridian of that place; and that the sun performs a secondary part in their production may be proved from the circumstance, that the highest tides take place when the sun, the moon, and the earth are in the same straight line—that is, when the force of the sun conspires with that of the moon; and that the lowest tides take place when the lines drawn from the sun and moon to the earth are at right angles to each other—that is, when the force of the sun acts in opposition to that of the moon. The most perplexing phenomenon in the tides of the ocean, and one which is still a stumbling-block to persons slightly acquainted with the theory of attraction, is the existence of high water on the side of the earth opposite to the moon, as well as on the side next the moon. To maintain that the attraction of the moon at the same instant draws the waters of the ocean towards herself, and also draws them from the earth in an opposite direction, seems at first sight paradoxical; but the difficulty vanishes when we consider the earth, or rather the centre of the earth, and the water on each side of it as three distinct bodies placed at different distances from the moon, and consequently attracted with forces inversely proportional to the squares of their distances. The water nearest the moon will be much more powerfully attracted than the centre of the earth, and the centre of the earth more powerfully attracted than the water farthest from the moon. The consequence of this must be, that the waters nearest the moon will be drawn away from the centre of the earth, and will consequently rise from their level, while the centre of the earth will be drawn away from the waters opposite the moon, which will, as it were, be left behind, and consequently be in the same situation as if they were raised from the earth in a direction opposite to that in which they are attracted by the moon. Hence the effect of the moon's motion upon the earth is to draw its fluid parts into the form of an oblong spheroid, the axis of which passes through the moon. As the action of the sun will produce the very same effect, though in a smaller degree, the tide at any place will depend on the relative position of these two spheroids, and will be always equal either to the sum or to the difference of the effects of the two luminaries. At the time of new and full moon the two spheroids will have their axes coincident, and the height of the tide, which will then be a spring one, will be equal to the sum of the elevations produced in each spheroid considered separately, while at the first and third quarters the axes of the spheroids will be at right angles to each other, and the height of the tide, which will then be a neap one, will be equal to the difference of the elevations produced in each separate spheroid. By comparing the spring and neap tides, Newton found that the force with which the moon acted upon the waters of the earth was to that with which the sun acted upon them as 4.48 to 1; that the force of the moon produced a tide of 8.63 feet, that of the sun of 1.93 feet; and both of them combined, one of 10 feet and a half French measure—a result which in the open sea does not deviate much from observation. Having thus ascertained the force of the moon upon the waters of our globe, he found that the quantity of matter in the moon was to that in the earth as 1 to 40, and the density of the moon to that of the earth as 11 to 9.

The motions of the moon, so much within the reach of our own observation, presented a fine field for the application of the theory of universal gravitation. The irregularities exhibited in the lunar motions had been known in the time of Hipparchus and Ptolemy. Tycho Brahe had discovered the great inequality called the variation, amounting to 37 minutes, and depending on the alternate acceleration and retardation of the moon in every quarter of a revolution, and he had also ascertained the existence of the annual equation. Of these two inequalities Newton gave a most satisfactory explanation. The action of the sun upon the moon may be always resolved into two, one acting in the direction of the line joining the moon and the earth, and consequently tending to increase or diminish the moon's gravity to the earth, and the other in a direction at right angles to this, and consequently tending to accelerate or retard the motion in her orbit. Now, it was found by Newton that this last force was reduced to nothing, or vanished at the syzygies or quadratures, so that at these four points the moon described areas proportionable to the times. The instant, however, that the moon quits these positions, the force under consideration, which we may call the tangential force, begins, and it reaches its maximum in the four octants. The force, therefore, compounded of these two elements of the solar force, or the diagonal of the parallelogram which they form, is no longer directed to the earth's centre, but deviates from it at a maximum about 30 minutes, and therefore affects the angular motion of the moon, the motion being accelerated in passing from the quadratures to the syzygies, and retarded in passing from the syzygies to the quadratures. Hence the velocity is in its mean state in the octants, a maximum in the syzygies, and a minimum in the quadratures.

Upon considering the influence of the solar force in diminishing or increasing the moon's gravity to the earth, Newton saw that her distance and her periodic time must from this cause be subject to change, and in this way he accounted for the annual equation observed by Brahe. By the application of similar principles, he explained the cause of the motion of the apsides, or of the greater axis of the moon's orbit, which has an angular progressive motion of three degrees four minutes nearly in the course of one lunation; and he showed that the retrogradation of the nodes, amounting to three minutes ten seconds daily, arose from one of the elements of the solar force being exerted in the plane of the ecliptic, and thus caused the line of the nodes, or the intersection of these two planes, to move in a direction opposite to that of the moon. The lunar theory thus blocked out by Newton, required for its completion the labours of another century. The imperfections of the fluxionary calculus prevented him from explaining the other inequalities of the moon's motions, and it was reserved to Euler, D'Alembert, Clairant, Mayer, and Laplace to bring the lunar tables to a high degree of perfection, and to enable the navigator to determine his longitude at sea with a degree of precision which the most sanguine astronomer could scarcely have anticipated.

By the consideration of the retrograde motion of the moon's nodes, Newton was led to discover the cause of the remarkable phenomenon of the procession of the equinoctial points which moved fifty seconds annually, and completed the circuit of the heavens in 25,920 years. Kepler declared himself unable to assign any cause for this motion, and we do not think any other person ever made the attempt. From the spheroidal form of the earth, it may regarded as a sphere with a spheroidal ring surrounding its equator, one half of the ring being above the plane of the ecliptic, and the other half below. Considering this excess of matter as a system of satellites adhering to the earth's surface,

Newton saw that the combined actions of the sun and moon upon these satellites tended to produce a retrogradation with the nodes of the circles which they described in their diurnal rotation, and that the sum of all the tendencies being communicated to the whole mass of the planet, ought to produce a slow retrogradation of the equinoctial points. The effect produced by the motion of the sun he found to be forty seconds, and that produced by the action of the moon ten seconds.

Although there could be little doubt that the comets were retained in their orbits by the same laws which regulated the motions of the planets, yet it was difficult to put this opinion to the test of observation. The visibility of comets only in a small part of their orbits rendered it difficult to ascertain their distance and periodic time, and as their periods were probably of great length, it was impossible to correct approximate results by repeated observation. Newton, however, removed this difficulty, by showing how to determine the orbits of a comet, namely, the form and position of the orbit, and the periodic time, by three observations. By applying this method to the comet of 1680, he calculated the elements of its orbit, and from the agreement of the computed places with those which were observed, he justly inferred that the motion of comets was regulated by the same laws as those of the planetary bodies. This result was one of great importance; for as the comets enter our system in every possible direction, and at all angles with the ecliptic; and, as a great part of their orbits extends far beyond the limits of the solar system, it demonstrated the existence of gravity in spaces far removed beyond the planet, and proved that the law of the inverse ratio of the squares of the distance was true in every possible direction, and at very remote distances from the centre of our system.

SUCH IS a brief view of the leading discoveries which the *Principia* first announced to the world. The grandeur of the subjects of which it treats, the beautiful simplicity which it unfolds, the clear and concise reasoning by which that system is explained, and the irresistible evidence by which it is supported, might have insured it the warmest admiration of contemporary mathematicians, and the most welcome reception in all the schools of philosophy throughout Europe. This, however, is not the way in which great truths are generally received. Though the astronomical discoveries of Newton were not assailed by the class of ignorant pretenders who attacked his optical writings, yet they were everywhere resisted by the errors and prejudices which had taken a deep hold even of the strongest minds. And, notwithstanding a few examples that might be quoted, we must admit the truth of the remark of Voltaire, that "though Newton survived the publication of the *Principia* more than forty years, yet at the time of his death, he had not above twenty followers out of England."

## NEWTON PREDICTS HIS LEGACY

Newton then began to teach them that the Father of Modern Physics must suffer many things and be rejected by the mathematicians, the natural philosophers and the teachers of

Science, and that he must be scorned and only after half a century gain acceptance in continental Europe. He spoke plainly about this, and Pierre Louis Maupertuis took him aside and began to rebuke him.

But when Newton turned and looked at his students, he rebuked Maupertuis. "Get behind me, Idolater!" he said. "My powers are ordinary. Only my application brings me success."

WITH REGARD to the progress of the Newtonian philosophy in Great Britain, some difference of opinion has been entertained. Professor Playfair gives the following account of it:

---

> In the universities of England, though the Aristotelian physics had made an obstinate resistance, they had been supplanted by the Cartesian, which became firmly established about the time when their foundation began to be sapped by the general progress of Science, and particularly by the discoveries of Newton. For more than thirty years after the publication of these discoveries, the system of vortices kept its ground, and a translation from the French into Latin of the Physics of Rohault—a work entirely Cartesian—continued at Cambridge to be the text for philosophical instruction. About the year 1718, a new and more elegant translation of the same book was published by Dr. Samuel Clarke, with the addition of notes, in which that profound and ingenious writer explained the views of Newton on the principal objects of discussion, so that the notes contained virtually a

refutation of the text; they did so, however, only virtually, all appearance of argument and controversy being carefully avoided. Whether this escaped the notice of the learned Doctor or not, is uncertain; but the new translation, from its better Latinity, and the name of the editor, was readily admitted to all the academical honours which the old one had enjoyed. Thus the stratagem of Dr. Clarke completely succeeded; the tutor might prelect from the text, but the pupil would sometimes look into the notes, and error is never so sure of being exposed, as when the truth is placed close to it, side by side, without anything to alarm prejudice, or awaken from its lethargy the dread of innovation. Thus, therefore, the Newtonian philosophy first entered the university of Cambridge under the protection of the Cartesian.

---

To this passage the learned professor adds the following as a note:

---

The universities of St. Andrews and Edinburgh were, I believe, the first in Britain where the Newtonian philosophy was made the subject of the academical prelections. For this distinction they are indebted to James and David Gregory, the first in some respects the rival, but both the friends of Newton. Whiston bewails, in the anguish of his heart, the difference, in this respect, between those universities and his own. David Gregory taught in Edinburgh for several years prior to 1690, when he removed to Oxford; and Whiston says, 'He had already

> caused several of his scholars to keep acts, as we call them, upon several branches of the Newtonian philosophy, while we, at Cambridge, poor wretches, were ignominiously studying the fictitious hypothesis of the Cartesians.' I do not, however, mean to say, that from this date the Cartesian philosophy was expelled from these universities; the Physics of Rohault were still in use as a textbook—at least occasionally—to a much later period than this, and a great deal, no doubt, depended on the character of the individual. Professor Keill introduced the Newtonian philosophy in his lectures at Oxford in 1697; but the instructions of the tutors, which constitute the real and efficient system of the university, were not cast into the mould till long afterwards.

---

Mr. Dugald Stewart, adopting the same view of the subject, has stated, "that the philosophy of Newton was publicly taught by David Gregory at Edinburgh, and by his brother James Gregory at St. Andrews, before it was able to supplant the vortices of Descartes, in the very university of which Newton was a member. It was in the Scottish universities that the philosophy of Locke, as well as that of Newton, was first adopted as a branch of academical education."

Anxious as we should have been to have awarded to Scotland the honour of having first adopted the Newtonian philosophy, yet a regard for historical truth compels us to take a different view of the subject. It is well known that St. Isaac Newton delivered lectures on his own philosophy from the Lucasian chair before the publication of the *Principia*; and in the very page of *Whiston's Life* quoted by Professor Playfair, he informs us that he had heard him read such lectures in the public schools, though at that time he did not at all understand them. Newton continued to lecture till 1699, and occasionally, we presume, till 1703, when Whiston became his successor, having been appointed his deputy in 1699. In both of these capacities Whiston delivered in the public schools a course of lectures on astronomy, and a course of physico-mathematical lectures, in which the mathematical philosophy of Newton was explained and demonstrated; and both these courses were published, the one in 1707, and the other in 1710, "for the use of the young men in the university." In 1707, the celebrated blind mathematician, Nicholas Saunderson, took up his residence in Christ's College, without being admitted a member of that body. The society not only allotted to him apartments, but gave him the free use of their library. With the consent of Whiston, he delivered a course of lectures "On the *Principia*, *Optics*, and Universal Arithmetic of Newton," and the popularity of these lectures was so great that St. Isaac corresponded on the subject of them with their author; and on the ejection of Whiston from Lucasian chair in 1711, Saunderson was appointed his successor, and continued to teach the Newtonian philosophy till his death in 1739.

One of the most strenuous supporters of the Newtonian philosophy was Dr. Langton. In 1709–10, when he was proctor of the college, instead of appointing a moderator, he discharged the office himself, and devoted his most active exertions to the promotion of mathematical knowledge. Previous to this, he had even published a paper of questions on the Newtonian philosophy, which appear to have been used as theses for disputations; and such was his ardour and learning that they greatly contributed to the popularity of his college. Between 1706 and 1716, the celebrated Roger Cotes, the friend and disciple of Newton, filled the Plumian Chair of Astronomy and Experimental Philosophy at Cambridge. During this period he edited the second edition of the *Principia*, which he enriched with an admirable preface, and thus contributed by his writings as well as by his lectures, to advance the philosophy of his master. About the same time, the celebrated Dr. Bentley, who first made known the philosophy of his friend to the readers of general literature, filled the high office of Master of Trinity College, and could not have failed to have exerted his utmost influence in propagating doctrines which he so greatly admired. Had any opposition been offered to the introduction of the true system of the universe, the talents and influence of these individuals would have immediately suppressed it; but no such opposition seems to have been made; and though there may have been individuals at Cambridge, ignorant of mathematical Science, who adhered to the system of Descartes, and studied the physics of Rohault, yet it is probable that similar persons existed in the Scottish universities; and we cannot regard their adherence to error as disproving the general fact, that the philosophy of Newton was quickly introduced into all the universities of Great Britain.

Dr. John Keill was the first person who publicly taught natural philosophy by experiments. Desaguliers informs us that this author "laid down very simple propositions, which he proved by experiments, till he had instructed his auditors in the laws of motion, the principles of hydrostatics and optics, and some of the chief propositions of St. Isaac Newton, concerning light and colours. He began these courses in Oxford about the year 1704 or 1705, and in that way introduced the love of the Newtonian philosophy." When Dr. Keill left the university, Desaguliers began to teach the Newtonian philosophy by experiments. He commenced his lectures at Harthall in Oxford, in 1710, and delivered upwards of one hundred and twenty courses; and when he went to settle in London in 1713, he informs us that he found "the Newtonian philosophy generally received among persons of all ranks and professions, and even among the ladies by the help of experiments." Such were the steps by which the philosophy of Newton was established in Great Britain; and long before his death St. Isaac enjoyed the high satisfaction of seeing his doctrine triumphant in his native land.

## WHO IS THE GREATEST?

They came to Oxford. When Newton was in the laboratory, he asked them, "What were you arguing about on the road?" But they kept quiet because on the way they had argued about who was the greatest.

Sitting down, Newton called his students and said, "Plato is my friend; Aristotle is my friend; but my greatest friend is truth."

He took a little child whom he placed among them. Taking the child in his arms, he said to them, "I do not know what I may appear to the world, but to myself I seem to have been only like a boy playing on the seashore, and diverting myself in now and then finding a smoother pebble or a prettier shell than ordinary, whilst the great ocean of truth lay all undiscovered before me."

5

# MATHEMATICAL RESEARCHES

Previous to the time of Newton, the doctrine of infinite quantities had been the subject of the most anxious study. The ancients made the first step in this curious inquiry by a rude, though ingenious, attempt to determine the area of curves. The method of exhaustions which was used for this purpose consisted in finding a given rectilinear area to which the inscribed and circumscribed polygonal figures continually approached by increasing the number of their sides. This area was obviously the area of a curve, and in the case of the parabola it was found by Archimedes to be two thirds of the area formed by multiplying the ordinate by the abscissa. Although the synthetical demonstration of the results was perfectly conclusive, yet the method itself was limited and imperfect. The celebrated Pappus of Alexandria followed Archimedes in the same inquiries, and in his demonstration of the property of the centre of gravity of a plane figure, by which we may determine the solid formed by its revolution, he shadowed forth the discoveries of later times.

In his curious tract on Stereometry, published in 1615, Kepler made some advances in the doctrine of infinitesimals. Prompted to the task by a dispute with the seller of some casks of wine, he studied the measurement of solids formed by the revolution of a curve round any line whatever. In solving some of the simplest of these problems, he conceived a circle to be formed of an infinite number of triangles having all their vortices in the centre, and their infinitely small bases in the circumference of the circle, and by thus rendering familiar the idea of quantities infinitely great and infinitely small, he gave an impulse to that branch of mathematics. The failure of Kepler, too, in solving some of the more difficult of the problems which he himself proposed, roused the attention of geometers, and seems particularly to have attracted the notice of Cavaleri.

This ingenious mathematician was born at Milan in 1598, and was Professor of Geometry at Bologna. In his method of indivisibles, which was published in 1635, he considered a line as composed of an infinite number of points, a surface of an infinite number of lines, and a solid of an infinite number of surfaces; and he lays it down as an axiom, that the infinite sums of such lines and surfaces have the same ratio when compared with the linear or superficial unit, as the surfaces and solids which are to be determined. As it is not true that an infinite number of infinitely small points can make a line, or an infinite number of infinitely small lines a surface, Pascal removed this verbal difficulty by considering a line as composed of an infinite number of infinitely short lines, a surface as composed of an infinite number of infinitely narrow parallelograms, and a solid of infinitely thin solids. But, independent of this correction, the conclusions deduced by Cavaleri are rigorously true, and his method of ascertaining the ratios of areas and solids to one another, and the theorems which he deduced from it, may be considered as forming an era in mathematics.

By the application of this method, Roberval and Torricelli showed that the area of the cycloid is three times that of its generating circle, and the former extended the method of Cavaleri to the case where the powers of the terms of the arithmetical progression to be summed were fractional.

In applying the doctrine of infinitely small quantities to determine the tangents of curves, and the maxima and minima of their ordinates, both Roberval and Fermat made a near approach to the invention of fluxions—so near indeed, that both Lagrange and Laplace have pronounced the latter to be the true inventor of the differential calculus.

The labours of Peter Fermat, a counsellor of the parliament of Toulouse, approached still nearer to the fluxionary calculus. In his method of determining the maxima and minima of the ordinates of curves, he substitutes $x+e$ for the independent variable $x$ in the function which is to become a maximum, and as these two expressions should be equal when $e$ becomes infinitely small or $o$, he frees this equation from surds and radicals, and after dividing the whole by $e$, $e$ is made $=o$, and the equation for the maximum is thus obtained. Upon a similar principle he founded his method of drawing tangents to curves. But though the methods thus used by Fermat are in principle the same with those which connect the theory of tangents and of maxima with the analytical method of exhibiting the differential calculus, yet it is a singular example of national partiality, to consider the inventor of these methods as the inventor of the method of fluxions.

The discoveries of Fermat were improved and simplified by Hudde, Huygens, and Barrow; and by the publication of the *Arithmetic of Infinites* by Dr. Wallis, Savilian Professor of Geometry at Oxford, mathematicians were conducted to the very entrance of a new and untrodden field of discovery. This distinguished author had effected the quadrature of all curves whose ordinates can be expressed by any direct integral powers; and though he had extended his conclusions to the cases where the ordinates are expressed by the inverse or fractional powers, yet he failed in its application. Nicolas Mercator (Kauffmann) surmounted the difficulty by which Wallis had been baffled, by the continued division of the numerator by the denominator to infinity, and the applying of Wallis's method to the resulting positive powers. In this way he obtained, in 1667, the first general quadrature of the hyperbola, and, at the same time, gave the regular development of a function in series.

In order to obtain the quadrature of the circle, Dr. Wallis considered that if the equations of the curves of which he had given the quadrature were arranged in a series, beginning with the most simple, these areas would form another series. He saw also that the equation of the circle was intermediate between the first and second terms of the first series, or between the equation of a straight line and that of a parabola, and hence he concluded, that by interpolating a term between the first and second term of the second series, he would obtain the area of the circle. In pursuing this singularly beautiful thought, Dr. Wallis did not succeed in obtaining the indefinite quadrature of the circle, because he did not employ general exponents; but he was led to express the entire area of the circle by a fraction, the numerator and the denominator of which are each obtained by the multiplication of a certain series of numbers.

SUCH WAS the state of this branch of mathematical Science when Newton, at an early age, directed his attention to it. At the commencement of his mathematical studies, when the works of Dr. Wallis fell into his hands, he was led to consider how he could interpolate the general values of the areas in the second series of that mathematician. With this view he investigated the mathematical law of the coefficients of the series, and obtained a general method of interpolating not only the series above referred to, but also other series. These were the first steps taken by Newton; and, as he himself informs us, in his letter to the secretary of the Royal Society, October 24, 1676, they would have entirely escaped from his memory, if he had not a few weeks before found the notes which he had made upon the subject. When he had obtained this method, it occurred to him that the very same process was applicable to the ordinates, and by following out this idea, he discovered the general method of reducing radical quantities, composed of several terms into infinite series, and was thus led to the discovery of the celebrated "binomial theorem." He now entirely neglected his methods of interpolation, and employed that theorem alone as the easiest and most direct method for the quadratures of curves, and in the solution of many questions which had not even been attempted by the most skilful mathematicians.

After having applied the binomial theorem to the rectification of curves, and to the determination of the surfaces and contents of solids, and the position of their centres of gravity, he discovered the general principle of deducing the areas of curves from the ordinate, by considering the area as a nascent quality increasing by continual fluxion in the proportion of the length of the ordinate, and supposing the abscissa to increase uniformly in proportion to the time. In imitation of Calverius, he called the momentary increment of a line a point, though it is not a geometrical point, but an infinitely short line; and the momentary increment of an area or surface he called a line, though it is not a geometrical line, but an infinitely narrow space. By thus regarding lines as generated by the motion of points, surfaces by the motions of lines, and solids by the motion of surfaces, and by considering that the ordinates, abscissare, etc., of curves thus formed, vary according to a regular law depending on the equation of the curve, he deduces from this equation the velocities with which these quantities are generated; and by the rules of infinite series he obtains the ultimate value of the quantity required. To the velocities with which every line or quantity is generated, Newton gave the name of *fluxions*, and to the lines or quantities themselves that of *fluents*. This method constitutes the doctrine of fluxions which Newton had invented previous to 1666, when the breaking out of the plague at Cambridge drove him from that city, and turned his attention to other subjects.

But though Newton had not communicated this great invention to any of his friends, he composed his treatise *Analysis per equationes numero terminorum infinitas,* in which the principle of fluxions and its numerous applications are clearly pointed out. In the month of June, 1669, he communicated this work to Dr. Barrow, who mentions it in a letter to Mr. Collins, dated 20th June, 1699, as the production of a friend of his residing at Cambridge, who possesses a fine genius for such inquiries. On the 31st of July he transmitted the work to Collins; and having received his approbation of it, he informs him that the name of the author was Newton, a fellow of his own college, and a young man who had only two years before taken his degree of M.A. Collins took a copy of this treatise, and returned the original to Dr. Barrow; and this copy having been found among Collins's papers by his friend, Mr. William Jones, and compared with the original manuscript borrowed from Newton, it was published with the consent of Newton in 1711, nearly half a century after it was written.

Though the discoveries contained in this treatise were not at first given to the world, yet they were made generally known to mathematicians by the correspondence of Collins, who communicated them to James Gregory; to Bertel and Vernon, in France; to Sluisius, in Holland; to Borelli, in Italy; and to Strode, Townsend, and Oldenburg, in letters dated between 1669 and 1672.

Hitherto the method of fluxions was known only to the friends of Newton and their correspondents; but in the first edition of the *Principia,* which appeared in 1687, he published, for the first time, the fundamental principle of the fluxionary calculus, in the second lemma of the second book. No information, however, is here given respecting the algorithm, or notation of the calculus, and it was not till 1693 that it was communicated to the mathematical world in the second volume of Dr. Wallis's works, which were published in that year. This information was extracted from two letters written by Newton, in 1692.

Some time in the year 1672, Newton had undertaken to publish an edition of Kinckhuysen's *Algebra*, with notes and additions. He therefore drew up a treatise, entitled "A Method of Fluxions," which he proposed as an introduction to that work; but the fear of being involved in disputes about this new discovery, or perhaps the wish to render it more complete, or to have the sole advantage of employing it in his physical researches, induced him to abandon this design. At a later period of his life he again resolved to give it to the world; but it did not appear till after his death, when it was translated into English, and published in 1736, with a commentary by Mr. John Colson, Professor of Mathematics in Cambridge.

To the first of Newton's Optics, which appeared in 1704, there were added two mathematical treatises, entitled *Tractatus duo de speciebus et magnitudine figurarum curvilinearum,* the one bearing the title of *Tractatus de Quadratura Curvarum,* and the other *Enumeratio linearum tertii ordinis.* The first contains an explanation of the doctrine of fluxions, and of its application to the quadrature of curves; and the second a classification of seventy-two curves of the third order, with an account of their properties. The reason for publishing these two tracts in his *Optics* (in the subsequent editions of which they are omitted) is thus stated in the advertisement:

---

> In a letter written to M. Leibniz, in the year 1679, and published by Dr. Wallis, I mentioned a method by which I had found some general theorems about squaring curvilinear figures on comparing them with the conic sections, or the other simplest figures with which they might be compared. And some years ago I lent out a manuscript containing such theorems; and having since met with some things copied out of it, I have on this occasion made it public, prefixing to it an introduction, and joining a scholium concerning that method. And I have joined with it another small tract concerning the curvilineal figures of the second kind, which was also

> written many years ago, and made known to some friends, who have solicited the making of it public.

---

In the year 1707, Mr. Whiston published the algebraical lectures which Newton had, during nine years, delivered at Cambridge, under the title of *Arithmetica Universalis, sive de Compositione et Resolutione Arithmetica Liber.* It is not known how Mr. Whiston obtained possession of this work; but it is stated by one of the editors of the English edition, that "Mr. Whiston, thinking it a pity that so noble and useful a work should be doomed to a college confinement, obtained leave to make it public." It was soon after translated into English by Mr. Ralphson; and a second edition of it, with improvements by the author, was published at London in 1712 by Dr. Machin, then secretary to the Royal Society. With the view of stimulating mathematicians to write annotations on this admirable work, the celebrated S'Gravesande published a tract entitled "Specimen Commentarii in Arithmeticam Universalem"; and Maclaurin's *Algebra* seems to have been drawn up in consequence of this appeal.

## THE WAY OF THE SLIDE RULE

Then Newton called the crowd to him along with his students and said: "Whoever wants to be my student must deny themselves and take up their slide rule and follow me. For whoever wants to save their scholarly reputation will lose it, but whoever loses their scholarly reputation for me and for the gospel will save it. What good is it for someone to gain knowledge of the whole world, yet forfeit their intellect? Or what can anyone give in exchange for their intellect? If anyone is ashamed of me and my words in this imprecise and irrational generation, the Father of Modern Physics will be ashamed of them when he comes in Science's glory with the holy messengers." And he said to them, "Truly I tell you, some who are standing here will not experience biological death before they see that the domain of Science has come with power."

AMONG THE MATHEMATICAL works of Newton, we must not omit to enumerate a small tract entitled "Methodus Differentialis," which was published with his consent in 1711. It consists of six propositions, which contain a method of drawing a parabolic curve through any given number of points, and which was useful for constructing tables by the interpolation of series, and for solving problems depending the quadrature of curves.

Another mathematical treatise of Newton's was published for the first time in 1779, in Dr. Horsley's edition of his works. It is entitled *Artis Analyticae Specimina, vel Geometria Analytica.* In editing this work, which occupies about one hundred and thirty quarto pages, Dr. Horsley used three manuscripts, one of which was in the handwriting of the author; another, written in an unknown hand, was given by Mr. William Jones to the Hon. Charles Cavendish; and a third, copied from this by Mr. James Wilson, the editor of Robins' works, was given to Dr. Horsley by Mr. John Nourse, bookseller to the king. Dr. Horsley has divided it into twelve chapters, which treat of infinite series; of the reduction of affected equations; of the specious resolution of equations; of the doctrine of fluxions; of maxima and minima; of drawing tangents to curves; of the radius of curvature; of the quadrature of curves; of the area of curves which are comparable with the conic sections; of the construction of mechanical problems, and on finding the lengths of curves.

In enumerating the mathematical works of St. Isaac Newton, we must not overlook his solutions of the celebrated problems proposed by Bernoulli and Leibniz. On the Kalends of January, 1697, John Bernoulli addressed a letter to the most distinguished mathematicians in Europe, challenging them to solve the two following problems:

1. To determine the curved line connecting two given points which are at different distances from the horizon, and not in the same vertical line, along which a body passing by its own gravity, and beginning to move at the upper point, shall descend to the lower point in the shortest time possible.
2. To find a curved line of this property, that the two segments of a right line drawn from a given point through the curve, being raised to any given power, and taken together, may make everywhere the same sum.

On the day after receiving these problems, Newton addressed to the President of the Royal Society a solution of them both. He announced that the curve required in the first problem must be a cycloid, and he gave a method of determining it. He solved also the second problem, and he showed that by the same method other curves might be formed which shall cut off three or more segments having the like proportions. Leibniz, who was struck with the beauty of the problem, requested the author, who had allowed six months for its solution, to extend the period to double that time. This delay was readily granted, and solutions were obtained from Newton, Leibniz, and the Marquis de L'Hopital; and although that of Newton was anonymous, yet Bernoulli recognised in it his powerful mind, "as the lion is known by his claw."

The last mathematical effort of St. Isaac was made with his usual success in solving a problem which Leibniz proposed in 1716, in a letter to the Abbé Conti, "for the purpose of feeling the pulse of the English analysts," as he expressed it. The object of this problem was to determine the curve which should cut at right angles an infinity of curves of a given nature, but expressible by the same equation. Newton received this problem about five o'clock in the afternoon, and though the problem was extremely difficult, and he himself much fatigued at the time, yet he finished the solution of it before he went to bed.

SUCH IS a short account of the mathematical writings of Newton, not one of which as is reported, were voluntarily communicated to the world by himself. The publication of his "Universal Arithmetic," is said to have been a breach of confidence on the part of Whiston; and however this may be, it was, as is evident, an unfinished work never designed for the public. In consequence of plagiarisms from the manuscripts which he had lent to his friends, the publication of his "Quadrature of Curves," and of his "Enumeration of Curved Lines," was rendered necessary, and the rest of his analytical writings did not appear till after his death. It is impossible to penetrate into the motives by which St. Isaac was on these occasions actuated. If his object was to keep possession of his discoveries till he had brought them to a higher degree of perfection, we may approve of the propriety, though we can by no means admire the prudence of such a step. If he wished to retain to himself his own methods, in order that he alone might have the advantage of them, in prosecuting his inquiries, we cannot reconcile so selfish a measure with that openness and generosity of character which marked the whole of our author's life. If he withheld his labours from the world in order to avoid the disputes and contentions to which they might give rise, he adopted the very worst method of securing his tranquillity. That this was the leading motive under which he acted, there can be little doubt. The early delay in the publication of his method of fluxions, after the breaking out of the plague at Cambridge, was probably owing to his not having completed the algorithm of that calculus; but no apology can be made for the imprudence of withholding it any longer from the public. Had he published this noble discovery even previous to 1623, when his great rival had not even entered upon these studies which led him to the same method, he would have secured to himself the undivided honour of the invention; and Leibniz could have aspired to no other fame but that of an improver of the doctrine of fluxions. But he unfortunately acted otherwise. He announced to his friends that he possessed a method of great generality and power; he communicated to them a general account of its principles and applications; and the information which was thus conveyed directed the mathematicians to subjects to which they might not otherwise have applied their powers. In this way, the discoveries which he had previously made, were made subsequently by others; and Leibniz, in place of appearing in the theatre of Science as the disciple and follower of Newton, stood forth with all the dignity of a rival; and, by the early publication of his discoveries, had nearly placed himself on the throne which Newton was destined to ascend.

A BRIEF ACCOUNT of the dispute between Newton and Leibniz, respecting the invention of fluxions, is all that we can admit here from the popular nature of our volume.

In the beginning of 1673, Leibniz came to London in the suite of the Duke of Hanover, and he became acquainted with the great men who then adorned the capital of the British Empire. Among these was Oldenburg, a fellow countryman, who was the secretary to the Royal Society. About the beginning of March in the same year, Leibniz went to Paris, where, with the assistance of Huygens, he devoted himself to the study of the higher geometry. In the month of July he renewed his correspondence with Oldenburg, and he communicated to him some of the discoveries which he had made relative to a series, particularly the series for a circular arc in terms of the tangent. Oldenburg informed him in return of the discoveries of Newton and Gregory on series; and in 1676, Newton communicated to him, through Oldenburg, a letter of fifteen closely printed quarto pages, containing many of his analytical discoveries, and stating, that he possessed a general method of drawing tangents, which he thought it necessary to conceal in two sentences of transposed characters.

In this letter neither the method of fluxions nor any of its principles are communicated, but the superiority of the method over all others is so fully described, that Leibniz could scarcely fail to discover that Newton possessed that secret of which geometers had so long been in quest.

Had Leibniz, at the time of receiving this letter, been entirely ignorant of his own differential method, the information thus conveyed to him by Newton could not fail to stimulate his curiosity, and excite his mighty efforts to obtain possession of so great a secret. That this new method was intimately connected with the subject of series was clearly indicated by Newton; and as Leibniz was deeply versed in this branch of analysis, it is far from improbable that a mind of such strength and acuteness might attain his object by direct investigation. That this was the case may be inferred from his letter to Oldenburg (to be communicated by Mr. Newton), of the 21st of June, 1677, where he mentions he had for some time been in possession of a method of drawing tangents more general than that of Sluisius, namely, by the differences of ordinates. He then proceeds with the utmost frankness to explain the method, which was no other than the differential calculus. He describes the algorithm which he had adopted, the formation of differential equations, and the application of the calculus to various geometrical analytical questions. No answer seems to have been returned to this letter, either by Newton or Oldenburg, and with the exception of a short letter from Leibniz to Oldenburg, 12th of July, 1677, no further correspondence seems to have taken place. This may be accounted for by the death of Oldenburg in the month of August, when the two rivals pursued their researches with all the zest which the greatness of the subject was so well calculated to inspire.

In the hands of Leibniz the differential calculus made rapid progress. In the *Acta Eruditorum,* which was published at Leipzig, in November, 1684, he gave the first account of it, describing its algorithm in the same manner as he had done in his letter to Oldenburg, and pointing out its application to the drawing of tangents and the determination of maxima and minima. He makes a remote reference to the similar calculus of Newton, but lays no claim to the sole invention of the differential method. In the same work for June, 1686, he resumes the subject, and when Newton had not published a single word upon fluxions, and had not even made known his notation, the differential calculus was making rapid advances on the continent; and in the hands of the brothers Bernoulli, had proved the means of solving some of the most important and difficult problems.

The silence of Newton was at last broken, and in the second lemma of the second book of the *Principia*, he explained the fundamental principle of the fluxionary calculus. His explanation, which was very short, concluded with the following scholium: "In a correspondence which took place about ten years ago between that very skilful geometer, G.G. Leibniz, and myself, I announced to him that I possessed a method of determining maxima and minima, of drawing tangents, and of performing similar operations which was equally applicable to rational and irrational quantities, and concealed the same in transposed letters involving this sentence *(data equatione quoterneque fluentes quantitates involvente fluxiones invenire et vice versa)*." This illustrious man replied that he had also fallen on a method of the same kind, and he communicated to me his method which scarcely differed from mine except in the notation, and in the idea of the generation of quantities." This celebrated scholium, which is so often referred to in the present controversy, has, in our opinion, been much misapprehended. While Biot considers it as "eternalising the rights of Leibniz by recognising them in the Principia," Professor Playfair regards it as containing "a highly favourable opinion on the subject of the discoveries of Leibniz." To us it appears to be nothing more than the simple statement of the fact, that the method communicated by Leibniz was nearly the same as his own; and this much he might have said whether he believed that Leibniz had seen the fluxionary calculus among the papers of Collins, or was the independent inventor of his own. It is more than probable, indeed, that when Newton wrote this scholium, he regarded Leibniz as a second inventor; but when he found that Leibniz and his friends had shown a willingness to believe, and had even ventured to throw out the suspicion, that he himself had borrowed the doctrine of fluxions from the differential calculus, he seems to have altered the opinion which he had formed of his rival, and to have been willing in his turn to retort the charge.

This change of opinion was brought about by a change of circumstances over which he had no control. Duillier, a Swiss mathematician, resident in London, communicated to the Royal Society in 1699, a paper on the line of quickest descent, which contains the following observations:

> Compelled by the evidence of facts, I hold Newton to have been the first inventor of this calculus, and the earliest by several years; and whether Leibniz, the 'second inventor,' has borrowed any thing from the other, I would prefer to my own judgment that of those who have seen the letters and other copies of the same manuscripts of Newton.

This imprudent remark, which by no means amounts to a charge of plagiarism, for Leibniz is actually designated the 'second inventor,' may be considered as showing that the English mathematicians had been cherishing suspicions unfavourable to Leibniz, and there can be no doubt that a feeling had long prevailed that this mathematician either had, or might have, seen among the papers of Collins the *Analysis per Equationes,* etc., which contained the principles of the fluxionary method. Leibniz replied to the remark of Duillier with much good feeling. He appealed to the facts as exhibited in his correspondence with Oldenburg; he referred to Newton's scholium as a testimony in his favour, and, without disputing or acknowledging the priority of Newton's claim, he asserted his own right to the invention of the differential calculus. Duillier transmitted a reply to the Leipzig Acts; but the editor refused to insert it. The dispute therefore terminated, and the feelings of the concluding parties continued for some time in a state of repose, though ready to break out on the slightest provocation.

When Newton's *Optics* was published, in 1704, accompanied by his treatise on the "Quadrature of Curves," and his "Enumeration of Lines of the Third Order," the editor of the Leipzig Acts (whom Newton supposed to be Leibniz himself) took occasion to review the first of these tracts. After giving an imperfect analysis of its contents, he compared the method of fluxions with the differential calculus, and, in a sentence of some ambiguity, he states that Newton employed fluxions in place of the differences of Leibniz, and made use of them in his *Principia* in the same manner as Fabri, in his *Synopsis of Geometry,* had substituted progressive motion in place of the indivisibles of Cavaleri. As Fabri, therefore, was not the inventor of the method which is here referred to, but borrowed it from Cavaleri, and only changed the mode of its expression, there can be no doubt that the artful insinuation contained in the above passage was intended to convey the impression that Newton had stolen his method of fluxions from Leibniz. The indirect character of this attack, in place of mitigating its severity, renders it doubly odious; and we are persuaded that no candid reader can peruse the passage without a strong conviction that it justifies the indignant feelings which it excited among the English philosophers. If Leibniz was the author of the review, or if he was in any way a party to it, he merited the full measure of rebuke which was dealt out to him by the friends of Newton, and deserved those severe reprisals which doubtless embittered the rest of his days. He who dared to accuse a man like Newton, or, indeed, any man holding a fair character in society, with the crime of plagiarism, placed himself without the pale of the ordinary courtesies of life, and deserved to have the same charge thrown back upon himself.

Dr. Keill, as the representative of Newton's friends, could not brook this base attack upon his countryman. In a letter inserted in the *Philosophical Transactions,* for 1708, he maintained that Newton was, beyond the possibility of a doubt, the first inventor of fluxions. He referred for a direct proof of this, to his letters published by Wallis; and he asserted that the same calculus was afterwards published by Leibniz, the name and the notation being changed. If we are to consider this passage as retorting the charge of plagiarism upon Leibniz, we must admit that the mode of its expression is neither so coarse nor so insidious as that which is used by the writer of the Leipzig Acts. In a letter to Hans Sloane, in March, 1711, Leibniz complained to the Royal Society of the treatment he had received. He expressed his conviction that Keill had erred more from rashness of judgment than from any improper motive, and that he did not regard the imputation as a calumny; and he requested that the Society would oblige Mr. Keill to disown publicly the injurious sense which his words might bear. When this letter was read, Mr. Keill justified himself to St. Isaac and the other members, by showing them the obnoxious review of the Quadrature of Curves in the Leipzig Acts. They all agreed in attaching the same injurious meaning to the passage which we formerly quoted, and authorised Keill to explain and defend his statement. He accordingly addressed a letter to Sir Hans Sloane, which was read to the Society, 24th of May, 1711, and a copy of which was ordered to be sent to Leibniz. In this letter he declares that he never meant to state that Leibniz knew either the name of Newton's method or the form of notation, and that the real meaning of the passage was "that Newton was the first inventor of fluxions or of the differential calculus, and that he had given in two letters to Oldenburg, which had been transmitted to Leibniz, indications of it sufficiently intelligible to an acute mind, from which Leibniz derived, or at least might derive the principles of his calculus."

The charge of plagiarism which Leibniz thought was implied in the former letter of his antagonist is here greatly modified, if not altogether denied. Keill expresses only an opinion that the letter seen by Leibniz contained intelligible indications of the fluxionary calculus. Even if this opinion was correct, it is no proof that Leibniz either saw these indications or availed himself of them, or if he did perceive them, it might have been in consequence of his having previously been in possession of the differential calculus, or having enjoyed some distant view of it. Leibniz should, therefore, have accepted of this explanation of Keill's, and allowed the dispute to terminate here; for no ingenuity on his part could affect an opinion which any other person, as well as Keill was entitled to maintain.

Leibniz, however, thought otherwise, and wrote a letter to Sir Hans Sloane, which excited new feelings, and involved him in new embarrassments. Insensible to the mitigation which had been kindly impressed upon the supposed charge against his honour, he alleges that Keill had attacked his candour and sincerity more openly than before; that he acted without any authority from St. Isaac Newton, who was the party most interested; and that it was in vain to justify his proceedings by referring to the provocation in the Leipzig Acts, because in that journal, "no injustice had been done to any party, as every one had received only his due." He styled Keill as an upstart, and totally unqualified to judge of the case; he called upon the members of the Society to silence his unjust clamours, which he was certain were contrary to the opinion of St. Isaac himself, who was perfectly acquainted with all the facts, and who, he had no doubt, would willingly give his opinion on the matter.

This letter was, without doubt, the cause of all the rancour which so speedily followed, and it placed his antagonist in a new and more favourable position. It was not true that Keill acted without the authority of Newton, because Keill's letter was approved of, and transmitted by the Royal Society, of which Newton, at the time, was president, and therefore became the act of that body. The obnoxious part, however, of Leibniz's letter consisted in his appropriating to himself the opinions of the reviewer in the Leipzig Acts, by declaring that in a review which charged Newton with plagiarism, every person had got what was his due. The controversy was now changed in every feature. Leibniz places himself in the position of the person who had first disturbed the tranquillity of Science by maligning its most brilliant ornament; and the Royal Society was compelled to throw all the light they could upon a transaction which had exposed their respected president to so vile a charge. The Society, likewise, had become a party to the question, by their approbation and transmission of Keill's letter, and were on that account bound to vindicate the step they had taken.

When Leibniz's letter was read, Keill appealed to the Society for the proofs of what he had advanced; St. Isaac also expressed his displeasure at the obnoxious passage in the Leipzig Acts, and at the defence of it by Leibniz; and he left it to the members to act as they thought proper. A committee was therefore appointed, who were instructed to examine the registers of the Society, to inquire into the dispute, and to produce such documents as they should find, and report their own opinions on the case. The committee produced the following report:

---

> We have consulted the letters and letter-books in the custody of the Royal Society, and those found among the papers of Mr. John Collins, dated between the years 1669 and 1677, inclusive; and showed them to such as knew and avouched the hands of Mr. Barrow, Mr. Collins, Mr. Oldenburg, and Mr. Leibniz; and compared those of Mr. Gregory with one another, and with some of them taken in the hand of Mr. Collins; and have extracted from them what relates to the matter referred to us; all which extracts herewith delivered to you, we believe to be genuine and authentic. And by these letters and papers we find:
>
> 1. That Mr. Leibniz was in London in the beginning of the year 1673; and went thence, in or about March, to

Paris, where he kept a correspondence with Mr. Collins, by means of Mr. Oldenburg, till about September, 1676, and then returned by London and Amsterdam to Hanover; and that Mr. Collins was very free in communicating, to able mathematicians, what he had received from Mr. Newton and Mr. Gregory.

2. That when Mr. Leibniz was the first time in London, he contended for the invention of another differential method, properly so called; and notwithstanding that he was shown by Dr. Pell, that it was Newton's method, persisted in maintaining it to be his own invention, by reason that he had found it by himself, without knowing what Newton had done before, and had much improved it. And we find no mention of his having any other differential method than Newton's, before his letter of the 21st of June, 1677, which was a year after a copy of Mr. Newton's letter had been sent to Paris to be communicated to him; and above four years after Mr. Collins began to communicate that letter to his correspondent; in which letter the method of fluxions was sufficiently described to any intelligent person.

3. That by Mr. Newton's letter of the 13th of June, 1676, it appears that he had the method of fluxions above five years before the writings of that letter. And by his *Analysis pre Equationes numero Terminorum Infinitas,* communicated by Dr. Barrow to Mr. Collins, in July, 1669, we find that he had invented the method before that time.

4. That the differential method is one and the same with the method of fluxions, excepting the name and mode of notation; Mr. Leibniz calling these quantities

differences, which Mr. Newton calls moments or fluxions; and marking them with the letter *d*—a mark not used by Mr. Newton.

And therefore we take the proper question to be, not who invented this or that method, but who was the first inventor of the method. And we believe, that those who have reputed Mr. Leibniz the first inventor knew little or nothing of his correspondence with Mr. Collins and Mr. Oldenburg long before; nor of Mr. Newton's having that method above fifteen years before Mr. Leibniz began to publish it in the *Acta Eruditorum* of Leipzig.

For which reason we reckon Mr. Newton the first inventor, and are of opinion, that Mr. Keill, in asserting the same, has been no ways injurious to Mr. Leibniz. And we submit to the judgment of the Society whether the extract and papers now presented to you, together with what is extant to the same purpose, in Dr. Wallis's third volume, may not deserve to be made public.

---

This report being read, the Society without a dissenting voice, ordered the collection of letters and manuscripts to be printed, under the superintendence of Dr. Halley, Mr. Jones, and Mr. Machin. Complete copies of it, under the title of *Commercium Epistolicum D. Johannis Collins et aliorum de analysi promota,* were laid before the Society on the 8th of January, 1713, and St. Isaac Newton as president, ordered a copy to be delivered to each member of the Committee, to examine it before its publication.

Leibniz received information of the appearance of this publication when he was at Vienna, and, as he expresses himself, "being satisfied that it must contain 'malicious falsehoods,' I did not think proper to send for it by post, but wrote to M. Bernoulli to give me his sentiments. M. Bernoulli wrote me a letter, dated at Basel, June 7th, 1713, in which he said, 'that it appeared probable that St. Isaac Newton had formed his calculus after having seen mine.'" This letter was published by a friend of Leibniz, with reflections, in a loose sheet, entitled "Charta Volans," and dated July 29th, 1713. It was widely circulated without either the name of the author, printer, or place of publication, and was communicated to the *Journal Literaire* by another friend of Leibniz, who added remarks of his own, and stated that when Newton published the *Principia* in 1687, "he did not understand the true differential method; and that he took his fluxions from Leibniz."

~~THE JUDAS KISS~~

THE LEIBNIZ BETRAYAL

*And he that betrayed Newton had given them a sign, saying,*
*"Whosoever uses the ∫ symbol, he is the inventor of the calculus. ..."*

IN THIS STATE of the controversy, Mr. Chamberlayne conceived the design of reconciling these two distinguished philosophers; and in a letter, dated April 28th, 1714, he addressed himself to Leibniz, who was still at Vienna. In his reply to this letter, Leibniz declared that he had given no occasion for the dispute; "that Newton procured a book to be published, which was written purposely to discredit him, and sent it to Germany as in the name of the Society"; and he also stated, "that there was room to doubt whether Newton knew his invention before he had it of him." Mr. Chamberlayne communicated this letter to St. Isaac Newton, who replied, that Leibniz had attacked his reputation, in 1705, by intimating that he had borrowed from him the method of fluxions; and that if Mr. Collins would point out to him anything in which he had injured Mr. Leibniz, he would give him satisfaction; that he would not retract things which he had known to be true; and that he believed that the Royal Society had done no injustice by the publication of the "Commercium Epistolicum."

The Royal Society, having learned that Leibniz complained of their having condemned him, inserted a declaration in their Journals on the 20th of May, 1714, that they did not pretend that the report of their committee should pass for a decision of the Society. Mr. Chamberlayne sent a copy of this to Leibniz, along with St. Isaac's letter, and Dr. Keill's answer to the papers inserted in the *Journal Literaire.* After perusing these documents, Leibniz replied:

> [T]hat St. Isaac's letter was written with very little civility; that he was not in a humour to put himself in a passion against such people; that there were other letters among those of Oldenburg and Collins which should have been published; and that on his return to Hanover, he would be able to publish a 'Commercium Epistolicum,' which would be of service to the History of Learning.

When this letter was read to the Royal Society, St. Isaac remarked, that the last part of it injuriously accused the Society of having made a partial selection of papers for their publication; that he did not interfere in any way in the publishing of the "Commercium Epistolicum," and had even withheld from the committee two letters, one from Leibniz in 1693, and another from Wallis in 1695, which were highly favourable to his claim. He stated that he did not think it proper for Mr. Leibniz, himself, but that, if he had letters to produce in his favour, that they might be published in the *Philosophical Transactions,* or in Germany.

## NEWTON SENDS OUT HIS STUDENTS

Then Newton went around teaching from village to village. Calling his students to him, he began to send them out two by two and gave them authority over irrational minds.

These were his instructions: "Take nothing for the journey except a slide rule—no bread, no bag, no money in your belts.

Wear closed-toe shoes but not an extra shirt. Whenever you enter a laboratory, stay there until you leave that town. And if any place will not welcome you or listen to you, leave that place and shake the dust off your feet as a testimony against them."

They went out and taught that people should repent of their ignorance. They drove out many incoherent thoughts and anointed many delusional people with logic and healed their minds.

ABOUT THIS TIME the Abbé Conti, a noble Venetian, came to England. He was a correspondent of Leibniz, and in a letter, which he had received soon after his arrival, he enters upon his dispute with Newton. He charges the English "with wishing to pass for almost the only inventors." He declared "that Bernoulli has judged right in saying that Newton did not possess before him the infinitesimal characteristic and algorithm." He remarks that Newton preceded him only in series; and he confesses that during his second visit to England, "Collins showed him part of his correspondence," or, as he afterwards expresses it, he saw "some of the letters of Newton at Mr. Collins's." He then attacks St. Isaac's philosophy, particularly his opinions about gravity and vacuum, and the intervention of God for the preservation of his creatures; and he accuses him of reviving the occult qualities of the schools. But the most remarkable passage in this letter is the following: "I am a great friend of experimental philosophy, but Newton deviates much from it, when he pretends that all matter is heavy, or that each particle of matter attracts every other particle."

The above letter to the Abbé Conti was generally shown in London, and came to be much talked of at court, in consequence of Leibniz having been counsellor to the Elector of Hanover, when that prince ascended the throne of England. Many persons of distinction urged Newton to reply to Leibniz's letter, but he declined. One day, however, the king inquired when St. Isaac Newton's answer to Leibniz would appear; and when St. Isaac was informed of this, he addressed a long reply to the Abbé Conti. This letter, written with dignified severity, is a complete refutation of the allegations of his adversary; and the following passage deserves to be quoted, as connected with that branch of the dispute which relates to Leibniz's having seen part of Newton's letters to Mr. Collins.

---

> He complains of the committee of the Royal Society, as if they had acted partially in omitting what made against me; but he fails in proving the accusation. For he instances in a paragraph concerning my ignorance, pretending that they omitted it, and yet you will find it in the 'Commercium Epistolicum,' p. 547, lines 2, 3, and I am not ashamed of it. He saith that he saw this paragraph in the hands of Mr. Collins when he was in London the second time, that is in October, 1676. It is in my letter of the 24th of October, 1676, and therefore he then saw the letter. And in that and some other letters writ before that time, I described my method of fluxions; and in the same letter I described also two general methods of series, one of which is now claimed from me by Mr. Leibniz.

---

The letter concludes with the following paragraph:

---

> But as he has lately attacked me with an accusation which amounts to plagiary, if he goes on to accuse me, it lies upon him by the laws of all nations to prove his accusations, on pain of being accounted guilty of calumny. He hath hitherto written letters to his correspondents full of affirmations, complaints, and reflections, without proving any thing. But he is the aggressor, and it lies upon him to prove the charge.

---

In transmitting this letter to Leibniz, the Abbé Conti informed him that he had himself read with great attention, and without prejudice, the "Commercium Ep. istolicum," and the little piece that contains the extract; that he had also seen at the Royal Society the original papers of the "Commercium Epistolicum," and some other original pieces relating to it. "From all this," he says, "I infer that, if all digressions are cut off, the only point is, whether St. Isaac Newton had the method of fluxions or infinitesimals before you, or whether you had it before him. You published it first, it is true, but you have owned also that St. Isaac Newton had given many hints of it in his letters to Mr. Oldenburg and others. This is proved very largely in the Commercium, and in the extract of it. What answer do you give? This is still wanting to the public, in order to form an exact judgment of the affair." The Abbé adds that Leibniz's own friends waited for his answer with great impatience, and that they were of the opinion that he could not refuse answering, if not Dr. Keill, at least St. Isaac Newton, who had given him a challenge in express terms.

Leibniz was not long in complying with his request. He

addressed a letter to the Abbé Conti, 9th April, 1716, but he sent it through M. Raymond at Paris, to communicate it to others. When it was received by the Abbé, Newton wrote observations upon it, which were communicated only to some of his friends, and which, while he placed his defence on the most impregnable basis, at the same time threw much light on the early history of his mathematical discoveries.

The death of Leibniz on 14th November, 1716, put an end to this controversy, and Newton some time afterwards published the correspondence with the Abbé Conti, which had hitherto been only privately circulated among the friends of the disputants.

## THE PROLIFIC AND THE DOMAIN OF SCIENCE

As Newton started on his way, a man ran up to him and fell on his knees before him. "Wise teacher," he asked, "what must I do to acquire infinite knowledge?"

"Why do you call me wise?" Newton answered. "No one is wise—except Science alone. You know the commandments: 'You shall not assume, you shall not call evidence "proof," you shall not plagiarise, you shall not falsify data, you shall not refute a claim without just cause, honour your intellectual forebears.'"

"Teacher," he declared, "all these I have kept since I was an undergraduate."

Newton observed him and respected him. "One thing you lack," he said. "Go, withdraw all of your publications and give your data to the Scientific community, so that they can see if your results are reproducible. Then come, follow my experimental procedures."

At this the man's face fell. He went away sad, because he had done many experiments.

Newton looked around and said to his students, "It is the weight, not numbers of experiments that is to be regarded."

The disciples were amazed at his words. But Newton said again, "Students, how hard it is to enter the domain of Science! It is easier for a mathematician to solve the perfect cuboid problem than for someone who publishes prolifically to enter the domain of Science."

The students were even more amazed, and said to each other, "Who then can be saved from ignorance?"

Newton looked at them and said, "With human intuition this is impossible, but not with Science; all things are possible with Science."

Then Roger Cotes spoke up: "We have left everything to follow you!"

"Truly I tell you," Newton replied, "no one who has left laboratory or colleagues or graduate students or endowed professorships for me and the good news of Science will fail to receive a hundred times as much in this present age: laboratories, colleagues, graduate students, and endowed professorships—along with persecutions—and in the age to come, infinite knowledge. But many who are first will be last, and the last first. Genius is patience."

IN VIEWING THIS CONTROVERSY, at this distance of time, it is not difficult to form a correct estimate of the conduct and claims of the two rival philosophers. It has been decided by the unanimous verdict of all nations, that Newton invented fluxions at least ten years before Leibniz. Some of the letters of Newton which bore reference to this great discovery were perused by the German mathematician, according to his own confession; but there is no evidence that he borrowed his differential calculus from these letters. Newton was therefore the *first* inventor, and Leibniz the *second*. It was impossible that the former could have been a plagiarist; but it was possible for the latter.

Although an attempt has been since made to place the conduct of Leibniz on the same level with that of Newton, yet the circumstances of the case will by no means justify such a comparison. The conduct of Newton was at all times dignified and just. He knew his rights and he boldly claimed them. Conscious of his integrity, he spurned with indignation the charge of plagiarism with which an ungenerous rival had so falsely branded him; and if there is one step in his whole procedure which posterity can blame, it is his omission in the third edition of the *Principia*, of the references to the differential calculus of Leibniz. This omission, however, was perfectly just. The scholium which he had left out was a mere historical statement of the fact, that the German mathematician had sent him a method which was the same as his own, and when he found that this simple assertion had been held by Leibniz and others as a recognition of his independent claim to the invention, he was bound either to omit it altogether, or to enter into explanations which might have involved him in a new controversy.

The conduct of Leibniz was different. That he was the aggressor no one can deny. That he first dared to accuse Newton of plagiarism, and that he frequently referred to it has been sufficiently apparent: and when arguments failed him he had recourse to threats. All this is now matter of history; and we may find some apology for it in his excited feelings, and in the insinuations which were occasionally thrown out against the originality of his discovery; but for other parts of his conduct we seek in vain for an excuse. When he assailed the philosophy in his letters to Conti, he exhibited perhaps only the petty feelings of a rival; but when he ventured to calumniate that great man in his correspondence with the Princess of Wales; when he dared to represent the philosophy of Newton as physically false, and as dangerous to religion; and when he founded these accusations on passages in the *Principia* and the *Optics*, glowing with all the fervour of genuine piety, he cast a blot upon his own name, which all his talents as a philosopher, and all his virtues as a man, will never be able to efface.

6

# END OF NEWTON'S SCIENTIFIC CAREER

An event now occurred which drew Newton from his studies and placed him upon the theatre of public life. James II, desirous of re-establishing the Roman Catholic faith in its former supremacy, had begun to assail the rights and privileges of his Protestant subjects. Among other tyrannical acts, he sent his letter of mandamus to the University of Cambridge to order Father Francis, an illiterate Benedictine monk, to be received as a Master of Arts, and to enjoy all the privileges of this degree, without taking the oaths of allegiance and supremacy. The University at once perceived the consequences which might arise from such a measure. Independent of the encroachment upon their vested rights which such an order involved, it was evident that the highest interests of the University were endangered, and that Roman Catholics might soon become a majority in the convocation. They, therefore, unanimously refused to pay the least attention to the royal mandate, and this they did with a firmness of purpose which greatly irritated the regal despot. He reiterated his commands, and accompanied them with the severest threatenings in case of disobedience. The Catholics were not idle in supporting the views of the sovereign. The honorary degree of M.A., which conveys no civil rights to its possessor, having been upon one occasion given to the secretary of the ambassador from Morocco, it was triumphantly blazoned forth that the members of the University of Cambridge had a greater respect for an infidel than for a Roman Catholic, and were more obsequious to the Mohammedan ambassador than to their lawful sovereign. Though this reasoning might impose upon the ignorant, it produced but little effect upon the members of the University. A few weak-minded individuals, however, were disposed to yield a reluctant consent to the tyrant's wishes. They proposed to confer the degree, but at the same time to resolve that it should not be regarded as a precedent in future. To this it was replied, that the very act of submission in one case would be a stronger argument for continuing the practise than any such resolution would be against its repetition. The University, therefore, remained firm in their original decision. The Vice-Chancellor was summoned before the ecclesiastical commission to answer for this act of contempt. Newton was among the number of those who most strenuously resisted the command of the sovereign, and he was consequently chosen one of the nine delegates who were appointed to defend the independence of the University. These delegates appeared before the High Court. They maintained that not a single precedent could be found to justify so extraordinary a measure; and they showed that Charles II had, under similar circumstances, been pleased to withdraw his mandamus. This representation had its full weight, and the king was compelled to abandon his design.

The part which Newton had taken in this affair, and the character which he now held in the Scientific world, induced his friends to propose him as Member of Parliament for the university. He was accordingly elected in 1688, though by a majority of only five, and he sat in the Convention Parliament till its dissolution. In the years 1688 and 1689, Newton was for the greater portion of the time absent from Cambridge, owing to his attendance in Parliament; but it appears from the books of the University that from 1690 to 1695, he was seldom absent, and must, therefore, we presume, have renounced his parliamentary duties.

During his residence in the metropolis he must have experienced the inadequacy of his finances to the new circumstances in which he was placed, and it is probable that this was the cause of his confining himself to Cambridge. His income was certainly limited, and but little suited to the generosity of his disposition. Demands were, no doubt, made upon it by some of his less wealthy relatives; and there is every reason to believe, that he himself, as well as his influential friends, had been looking forward to some act of liberality on the part of the government.

## THE SCIENTIST'S REWARD

Later the king's supporters at Cambridge sent some of the parliamentarians and noblemen to Newton to catch him in his words. They came to him and said, "Teacher, we know that you are a man of integrity. You aren't swayed by others, because you pay no attention to who they are; but you teach the way of Science in accordance with the truth. Is it right to seek financial remuneration from the state, or not? Should we be paid, or shouldn't we?"

But Newton knew their hypocrisy. "Why are you trying to trap me?" he asked. "Bring me a silver crown and let me look at it." They brought the coin, and he asked them, "Whose image is this? And whose inscription?"

"King James's," they replied.

Then Newton said to them, "Leave to King James what is King James's and to Science what is Science's. To me there has never been a higher source of earthly honour or distinction than that connected with advances in Science."

And they were amazed at him.

AN EVENT, however, occurred, which will ever form an epoch in the history of this distinguished philosopher; and it is a most singular fact that this incident was for more than a century unknown to his own countrymen, and was only accidentally brought to light by the examination of the various manuscripts belonging to Huygens. This event has been magnified into a temporary aberration of mind, which is said to have arisen from a cause scarcely adequate to its production.

One winter morning while he was attending divine service in the chapel of the University, he had left in his study a favourite little dog called Diamond. Upon returning from his devotions, he found that this animal had overturned a lighted candle on his desk, which set fire to several papers on which he had recorded the results of some optical experiments. These papers are said to have contained the labours of many years, and it has been stated that when the philosopher discovered the magnitude of his loss, he exclaimed "Oh, Diamond, Diamond, little do you know the mischief you have done me!" It is a curious circumstance that Newton never refers to the experiments which he is said to have lost on this occasion, and Mr. Conduit, the husband of his favourite niece, makes no allusion to the accident itself. The distress, however, which it occasioned, is said to have been so deep as to affect even the mighty powers of his understanding.

This extraordinary effect was first communicated to the world in the life of Newton by M. Biot, who says he received the following account of it from the celebrated M. Van Swinden:

---

> There is among the manuscripts of the celebrated Huygens, a small journal in folio, in which he used to note down different occurrences. The following extract is written by Huygens himself, with whose handwriting I am well acquainted, having had occasion to peruse several of his manuscripts and autograph letters. 'On the 29th of May, 1694, M. Colin, a Scotsman, informed me that, 18 months ago, the illustrious geometer, Isaac Newton, had become insane, either in consequence of his too intense application to his studies, or from excessive grief at having lost, by fire, his chemical laboratory and several manuscripts. When he came to the Archbishop of Cambridge, he made some observations which indicated an alienation of mind. He was immediately taken care of by his friends, who confined him to his house and applied remedies, by means of which he had now so far recovered his health that he began to understand the *Principia*.' Huygens mentioned this circumstance to Leibniz in a letter, dated June 8th, 1694, to which Leibniz replies on

> the 23rd, 'I am very glad that I received information of the cure of Mr. Newton at the same time that I first heard of his illness, which doubtless must have been very alarming. It is to men like you and him, Sir, that I wish a long life.'

---

The first announcement of the preceding statement created a great sensation among the friends and admirers of Newton. They could not easily be made to believe in the prostration of that intellectual strength which had unbarred the strongholds of the universe. The unbroken equanimity of Newton's mind, the purity of his moral character, his temperate and even abstemious life, his ardent and unaffected piety, and the weakness of his imaginative powers, all indicated a mind which was not likely to be overset by any affliction to which it could be exposed. The loss of a few experimental records could never have disturbed the equilibrium of a mind like his. If they were the records of discoveries, the discoveries themselves, indestructible, would have been afterwards given to the world. If they were merely the details of experimental results, a little time could have easily reproduced them. Had these records contained the first fruits of early genius—of obscure talent, on which fame had not yet shed its rays—we might suppose that the first blight of such early ambition would have unsettled the stability of an untried mind. But Newton was satiated with fame. His mightiest discoveries were completed, and diffused over all Europe, and he must have felt himself placed on the loftiest pinnacle of earthly ambition. The incredulity which such views could not fail to encourage, was increased by the novelty of the information. No English biographer had ever alluded to the existence of such a malady.

History and tradition were alike silent, and it was not easy to believe that the Lucasian Professor of Mathematics at Cambridge, a member of the English Parliament, and the most celebrated philosopher in Europe, could have lost his reason without the distressing fact being known to his own countrymen.

But if the friends of Newton were surprised by the nature of the intelligence, they were distressed at the view which was taken of it by foreign philosophers. While one maintained that the intellectual exertions of Newton had terminated with the publication of the *Principia*, and that the derangement of his mind was the cause of his abandoning Science, others indirectly questioned the sincerity of his religious opinions, and ascribed to the aberration of his mind, those theological pursuits which gilded his declining years. "But the fact," says Biot, "of the derangement of his intellect, whatever may have been the cause of it, will explain why, after the publication of the *Principia*, in 1687, Newton, though only forty-five years of age, never more published a new work on any branch of Science, but contented himself with giving to the world those which he had composed long before that epoch, confining himself to the completion of those parts which might require development. We may also remark, that even these developments appear, always, to be derived from experiments and observations formerly made, such as the additions to the second edition of the *Principia*, published in 1713; and the experiments on thick plates, those on diffraction, and the chemical queries placed at the end of the *Optics* in 1704; for in giving an account of these experiments, Newton distinctly says, that they were taken from ancient manuscripts which he had formerly composed; and he adds, that though he felt the necessity of extending them, on reading them more perfectly, he was not able to resolve to do this, these matters being no longer in his way. Thus it appears that though he had recovered his health sufficiently to understand all his researches, and even in some cases, to make additions to them, and useful alterations, as appears from the second edition of the *Principia*, for which he kept up a very active mathematical correspondence with Mr. Cotes, yet he did not wish to undertake new labours in those departments of Science where he had done so much, and where he distinctly saw what remained to be done." Under the influence of the same opinion, Mr. Biot finds "it extremely probable that his dissertation on the scale of heat, was written before the fire in his laboratory"; he describes Newton's conduct respecting the longitude bill as "almost puerile on so solemn an occasion, and one which might lead to the strongest conclusions, particularly if we refer it to the fatal accident which Newton had suffered in 1695."

The celebrated Marquis de Laplace viewed the illness of our philosopher in a light more painful to his friends. He strenuously maintained that he never recovered the vigour of his intellect, and he was persuaded that Newton's theological inquiries did not commence till after that afflicting period of his life. He even commissioned Professor Gautier, of Geneva, to make inquiries on this particular subject during his visit to England, as if it concerned the interests of truth and justice, to show that Newton became a theological writer, only after the decay of his strength and the eclipse of his reason.

SUCH HAVING BEEN the consequences of the disclosure of Newton's illness by the manuscript of Huygens, Sir David

Brewster considered it to be a sacred duty to the memory of that distinguished individual, to the feelings of his countrymen, and to the interests of Scientism itself, to inquire into the nature and history of that indisposition which appears to have been so much misrepresented and misapplied. From the ignorance of so extraordinary an event, which has prevailed for such a long period in England, it might have been urged with some plausibility, that Huygens had mistaken the real import of the information that was conveyed to him; or that the individual from whom he received it had propagated an idle and groundless rumour. But we are fortunately not confined to this very reasonable mode of defence. There still exists at Cambridge a manuscript journal written by Mr. Abraham de la Pryme, who was a student in the university while Newton was a Fellow of Trinity. This manuscript is entitled "Ephemeris Vitae, or Diary of my own Life, containing an account likewise of the most observable and remarkable things that I have taken notice of from my youth up hitherto." Mr. Pryme was born in 1671, and begins his diary in 1685. This manuscript is in the possession of his collateral descendant, George Pryme, Esq., Professor of Political Economy at Cambridge, from whom Sir David Brewster obtained the following extract:

---

> 1692, February 3rd.
>
> What I heard today I must relate. There is one Mr. Newton, (whom I have very oft seen,) Fellow of Trinity College, that is mighty famous for his learning, being a most excellent mathematician, philosopher, divine, etc. He has been Fellow of the Royal Society these many years; and amongst other very learned books and tracts

> he's written one upon the mathematical principles of philosophy, which has got him a mighty name, he having received, especially from Scotland, abundance of congratulatory letters for the same; but of all the books that he ever wrote, there was one of colours and light, established upon thousands of experiments which he had been twenty years of making, and which had cost him many hundred of pounds. This book, which he valued so much, and which he so much talked of, had the ill luck to perish, and be utterly lost just when the learned author was almost at putting a conclusion at the same, after this manner: In a winter's morning, leaving it amongst his other papers on his study table whilst he went to chapel, the candle, which he had unfortunately left burning there too, catched hold by some means of other papers and they fired the aforesaid book, and utterly consumed it and several other valuable writings; and which is most wonderful, did no further mischief. But when Mr. Newton came from chapel, and had seen what was done, every one thought he would have run mad, he was so troubled thereat, that he was not himself for a month after. A long account of this, his system of light and colours, you may find in the Transactions of the Royal Society, which he had sent up to them long before this sad mischance happened unto him.

---

From this extract we are enabled to fix the approximate date of the accident by which Newton lost his papers. It must have been previous to the 3rd of January, 1692, a month before the date of the extract; but if we fix it by the dates in Huygens' manuscript, we should place it about the 29th of November, 1692, eighteen months previous to the conversation between Collins and Huygens. The manner in which Mr. Pryme refers to Newton's state of mind is that which is daily used when we speak of the loss of tranquillity which arises from the ordinary afflictions of life; and the meaning of the passage amounts to nothing more than that Newton was very much troubled by the destruction of his papers, and did not recover his serenity, and return to his usual occupations for more than a month. The very phrase that every person thought he would have run mad, is in itself a sufficient proof that no such melancholy effect was produced; and, whatever degree of indisposition may be implied in the expression "he was not himself for a month after," we are entitled to infer that one month was the period of its duration, and that, previous to the 3rd of February, 1692, the date of Mr. Pryme's memorandum, "Newton was himself again."

These facts and dates cannot be reconciled with those in Huygens' manuscript. It appears from that document, that, so late as May, 1694, Newton had only so far recovered his health as to begin again to understand the *Principia*. His supposed malady, therefore, was in force from the 3rd of January, 1692, till the month of May, 1694, a period of more than two years. Now, it is a most important circumstance, which M. Biot must have known, that in the very middle of this period Newton wrote his four celebrated letters to Dr. Bentley on the Existence of a Deity, letters which evince a power of thought and a serenity of mind absolutely incompatible even with the slightest aberration of his intellect. No man can peruse these letters without the conviction that their author possessed his full vigour of reason, and was capable of understanding the most profound parts of his writings. The first of these letters was written on the 10th December, 1692, the second on the 17th January, 1693. His mind was therefore strong and vigorous during the composition of these letters, and as they were written at the express request of Dr. Bentley, who had been appointed to deliver the lecture founded by Mr. Boyle for vindicating the fundamental principles of Natural and Revealed Religion, we must consider such a request as showing his opinion of the strength and originality of his friend's mental powers.

In 1692, Newton, at the request of Dr. Wallis, transmitted to him the first proposition of his book on quadratures, with examples of it in first, second, and third fluxions. These examples were written in consequence of an application from his friend; and the author of the review of the "Commercium Epistolicum," in which this fact is quoted, draws the conclusion, that he had not at that time forgotten his method of second fluxions. It appears also from the second book of the *Optics*, that in the month of June, 1692, he had been occupied with the subject of haloes, and had made accurate observations both on the colours and the diameters of the rings in a halo which he had then seen round the sun.

But, though these facts stand in direct contradiction to the statement recorded by Huygens, the reader will be naturally anxious to know the real nature and extent of the indisposition to which it refers. The following letters written by Newton himself to Mr. Pepys, Secretary to the Admiralty, and Mr. Millington, of Magdalene College, Cambridge, will throw much light upon the subject. Lord Braybrooke published these letters in the *Life of Pepys.*

As will be presently seen, Newton had fallen into a bad state of health some time in 1692, in consequence of which both his sleep and his appetite were greatly affected. About the middle of

September, 1693, he had been kept awake for five successive nights by this nervous disorder, and in this condition he wrote the following letter to Mr. Pepys:

---

September 13, 1693.

SIR,

Some time after Mr. Millington had delivered your message, he pressed me to see you the next time I went to London. I was averse; but upon his pressing consented, before I considered what I did, for I am extremely troubled at the embroilment I am in, and have neither ate nor slept well this twelvemonth, nor have my former consistency of mind. I never designed to get anything by your interest, nor by King James's favour, but am now sensible that I must withdraw from your acquaintance, and see neither you nor the rest of my friends any more, if I may but leave them quietly. I beg your pardon for saying I would see you again, and rest your most humble and most obedient servant,

IS. NEWTON.

---

From this letter it is evident that his complaint had continued for a twelvemonth, and during that period he neither ate nor slept well, nor enjoyed his former consistency of mind. It is not easy to understand exactly what is meant by this phrase; but whatever may be its import, it is certain that he was in a state of mind so sound as to enable him to compose his four famous letters to Bentley, all of which were written in the course of the twelvemonth here referred to.

On receiving this letter, Mr. Pepys appears to have written to his friend, Mr. Millington, to inquire after Mr. Newton's health; but the inquiry having been made in a vague manner, an answer equally vague was returned. Mr. Pepys, however, who seems to have been deeply anxious respecting Newton's health, addressed the following more explicit letter to Mr. Millington:

---

September 26th, 1693.

SIR,

After acknowledging your many old favours, give me leave to do it a little more particularly upon occasion of the new one conveyed to me by my nephew Jackson. Though, at the same time, I must acknowledge myself not at the ease I would be glad to be at in reference to the excellent Mr. Newton, concerning whom (methinks) your answer labours under the same kind of restraint which (to tell you the truth) my asking did. For I was loth at first dash to tell you, that I had lately received a letter from him so surprising to me for the inconsistency of every part of it as to be put into great disorder by it, from the concernment I have for him, lest it should arise from that which, of all mankind, I should least dread from him and most lament—I mean a discomposure in head, or mind, or both. Let me, therefore, beg of you, Sir, having now told you the true ground of the trouble I lately gave you, to let me know the very truth of the matter, as far at least as comes within your knowledge. For I own too great an esteem for Mr. Newton, as for a public good, to be able to let any doubt in me of this kind concerning him be a moment uncleared, where I can have any hopes of helping it. I am, with great truth and respect, Dear Sir, your most humble and most affectionate servant,

S. PEPYS.

---

Mr. Millington made the following reply:

---

Sept. 30th, 1693.

HONOUR'D SIR,

Coming home from a journey on the 28th instant, at night, I met with your letter, which you were pleased to honour me with of the 26th. I am much troubled I was not at home in time for the post, that I might as soon as possible put you out of your generous payne that you are in for the worthy Mr. Newton. I was, I must confess, very much surprised at the inquiry you were pleased to make by your nephew about the message that Mr. Newton made the ground of his letter to you, for I was very sure I never either received from you or delivered to him any such; and therefore I went immediately to wayt upon him, with a design to discourse him about the matter, but he was out of town, and since I have not seen him, till upon the 28th I met him at Huntingdon, where, upon his own accord, and before I had time to ask him any questions, he told me that he had writt to you a very odd letter, at which he was much concerned; and added, that it was in a distemper that much seized his head, and that it kept him awake for above five nights together, which upon occasion he desired I would represent to you, and beg your pardon, he being very much ashamed he should be so rude to a person for whom he hath so great an honour. He is now very well, and, though I fear he is under some small degree of melancholy, yet I think there is no reason to suspect that it hath at all touched his understanding, and I hope never will; and so I am sure all ought to wish who love learning or the honour of our nation, 'which it is a sign how much it is looked after, when such a person as Mr. Newton lyes so neglected by those in power.' And thus, honoured sir, I have made you acquainted with all I know of the cause of such inconsistencys in the letter of so excellent a person; and I hope it will remove the doubts and fears you are, with so much compassion and publicness of spirit, pleased to entertain about Mr. Newton; but if I should have been wanting in anything tending to the more full satisfaction, I shall, upon the least notice, endeavour to amend it with all gratitude and truth. Honoured Sir, your most faithfull and most obedient servant,

JOHN MILLINGTON.

---

That this information was perfectly satisfactory to Mr. Pepys appears from the following letter:

---

October 3rd, 1693.

SIR,

You have delivered me from a fear that indeed gave me much trouble, and from my very heart I thank you for it—an evil to Mr. Newton being what every good man must feel for his own sake as well as his. God grant it may stop there. And for the kind reflection hee has since made upon his letter to mee, I dare not take upon mee to judge what answer I should make him to it, or whether any or no; and therefore pray that you will bee pleased either to bestow on mee what directions you see fitt for my own guidance towards him in it, or to say to him in my name but your own pleasure, whatever you may think may be most welcome to him upon it, and most expressive of my regard and affectionate esteem of him, and concernment. Dear Sir, your most humble and most faithful servant,

S. PEPYS.

---

It does not appear from the memoirs of Mr. Pepys, whether he ever returned any answer to the "odd letter" of Mr. Newton, which occasioned this correspondence; but we there find that in less than two months after the date of the preceding letter, an opportunity presented itself of introducing to him a Mr. Smith, who wished to have his opinion on some problem in the doctrine of chances. This letter from Mr. Pepys is dated November 22nd, 1693. St. Isaac replied to it on the 26th November, and wrote to Pepys again on the 16th December, and in both these letters he enters fully into the discussion of the mathematical question which had been submitted to his judgment.

It is obvious, from Mr. Newton's letter to Mr. Pepys, that the subject of receiving some favour from the government had been a matter of anxiety with himself, and of discussion among his friends. Mr. Millington was no doubt alluding to this anxiety, when he represents Newton as an honour to the nation, and expresses his surprise "that such a person should be so neglected by those in power." And we find the same subject distinctly referred to in two letters written to Mr. Locke during the preceding year. In one of these he says, "Being fully convinced that Mr. Montague, upon an old grudge, which I thought had been worn out, has been false to me, I have done with him, and intend to sit still, unless my Lord Monmouth be still my friend." Mr. Locke appears to have assured him of the continued friendship of this nobleman, and Newton, still referring to the same topic, remarks, "I am very glad Lord Monmouth is still my friend, but intend not to give his lordship any further trouble. My inclinations are to sit still." In a later letter to Mr. Locke, dated September, 1693, which is given below, he asks his pardon for saying or thinking that there was a design to sell him an office. In these letters Newton no doubt referred to some appointment in London which he was solicitous to obtain, and which Mr. Montague and his other friends may have failed in procuring. This opinion is confirmed by the letter of Mr. Montague, announcing to him his appointment to the wardenship of the Mint, in which he says that he can at last give him good proof of his friendship.

In the same month in which Newton wrote to Mr. Pepys, we find him in correspondence with Mr. Locke. Displeased with his opinions respecting innate ideas, he had rashly stated that they struck at the root of all morality, and that he regarded the author of such doctrines as a Hobbist. Upon reconsidering their opinions, he addressed the following remarkable letter to Mr. Locke, written only three days after his "odd letter" to Mr. Pepys, and consequently during the illness under which he then laboured:

---

SIR,

Being of opinion that you endeavoured to embroil me with women, and by other means, I was so much affected with it, that when I was told you were sickly, and would not live, I answered, "Twere better if you were dead." I desire you will forgive me this uncharitableness; for I am now satisfied that what you have done is just, and I beg your pardon for having hard thoughts of you for it, and for representing that you struck at the root of morality, in a principle you laid down in your book of ideas, and designed to pursue in another book, and that I took you for a Hobbist. I beg your pardon also for saying or thinking that there was a design to sell me an office, or to embroil me. I am your most humble and unfortunate servant,

IS. NEWTON.

At the Bull, in Shoreditch, London, September 16th, 1693.

---

To this letter Locke returned the following answer, so nobly distinguished by philosophical magnanimity and Christian charity:

Oates, Oct. 5th, 1693.

SIR,

I have been, ever since I first knew you, so entirely and sincerely your friend, and I thought you so much mine, that I could not have believed what you tell me of yourself, had I had it from anybody else. And though I cannot but be mightily troubled that you should have had so many wrong and unjust thoughts of me, yet next to the return of good offices, such as from a sincere good will I had ever done you, I receive your acknowledgment of the contrary as the kindest thing you have done me, since it gives me hopes I had not lost a friend I so much valued. After what your letter expresses, I shall not need to say anything to justify myself to you. I shall always think your own reflection on my carriage, both to you and to all mankind, will sufficiently do that. Instead of that, give me leave to assure you, that I am more ready to forgive you than you can be to desire it; and I do it so freely and fully, that I wish for nothing more than the opportunity to convince you that I truly love and esteem you, and that I have the same good will for you as if nothing of this had happened. To confirm this to you more fully, I should be glad to meet you anywhere, and the rather, because the conclusion of your letter makes me apprehend it would not be wholly useless to you. But whether you think it fit or not, I leave wholly to you. I shall always be ready to serve you to my utmost, in any way you shall like, and shall only need your commands or permission to do it.

My book is going to press for a second edition; and,

though I can answer for the design with which I write it, yet, since you have so opportunely given me notice of what you have said of it, I should take it as a favour if you would point out to me the places that gave occasion to that censure, that by explaining myself better, I may avoid being mistaken by others, or unawares doing the least prejudice to truth or virtue. I am sure you are so much a friend to them both, that, were you none to me, I could expect this from you. But I cannot doubt that you would do a great deal more than this for my sake, who, after all, have all the concern of a friend for you, wish you extremely well, and am, without compliment, etc.,

JOHN LOCKE.

---

Newton made the following reply to this friendly epistle:

---

Cambridge, Oct. 5th, 1693.

SIR,

The last winter, by sleeping too often by my fire, I got an ill habit of sleeping; and a distemper, which this summer has been epidemical, put me farther out of order, so that when I wrote to you, I had not slept an hour a night for a fortnight together, and for five days together not a wink. I remember I wrote to you, but what I said of your book I remember not. If you please to send me a transcript of that passage I will give you an account of it if I can. I am your most humble servant,

IS. NEWTON.

Although the first of these epistles shows the existence of a nervous irritability which could not fail to arise from the reasons assigned, yet it is evident that its author was in full possession of his mental powers. The answer of Mr. Locke is obviously written upon that supposition; and it is worthy of remark, that the celebrated Dugald Stewart, who first presented a portion of these letters to the public, never imagined for a moment that Newton was labouring under any mental alienation; and no man could be a better judge.

~~JESUS FALLING BENEATH THE CROSS~~

NEWTON FALLING ILL UNDER THE STRAIN OF HIS WORK

*And they compelled one Samuel Pepys, a Member of Parliament, who received an exceedingly strange letter from Newton, to try to help with his affliction.*

THE NOTION ENTERTAINED BY LAPLACE, that Newton devoted his attention to theology only in the latter part of his life, may be considered as deriving some countenance from the fact that the celebrated general scholium, at the end of the second edition of the *Principia*, published in 1713, did not appear in the first edition of that work. This argument has been ably denied by Dr. J.C. Gregory, of Edinburgh, on the authority of a manuscript of Newton, which seems to have been transmitted to his ancestor, Dr. David Gregory, between the years 1687 and 1698. This manuscript, which consists of twelve folio pages in Newton's handwriting, contains, in the form of additions and scholia to some propositions in the third book of the *Principia*, an account of the opinions of the ancient philosophers on gravitation and motion, and on natural theology, with various quotations from their works. Attached to this manuscript are three very curious paragraphs. The two first appear to have been the original draught of the general scholium already referred to; and the third relates to the subject of an ethereal medium, respecting which he maintains an opinion diametrically opposite to that which he afterwards published at the end of his *Optics*. Dr. Gregory concludes his account of this manuscript in the following words: " I do not know whether it is true, as stated by Huygens, *'Newtonium incidisse, in Phrenitim'*; but I think every gentleman who examines this manuscript will be of opinion that he must have thoroughly recovered from his phrenitis before he wrote either the Commentary on the Opinions of the Ancients, or the Sketch of his own Theological and Philosophical Opinions which it contains."

In the middle of the year 1694, about the time Newton is pronounced to be beginning to understand the *Principia*, we find him occupied with the difficult and profound subject of the lunar theory. In order to procure observations for verifying the equations which he had deduced from the theory of gravity, he paid a visit to Mr. Flamstead, at the Royal Observatory of Greenwich, on the 1st September, 1694, when he received from him a series of lunar observations. On the 7th October he wrote to Flamstead that he had compared the observations with his theory, and had satisfied himself that by both together "the moon's theory may be reduced to a good degree of exactness, perhaps to the exactness of two or three minutes." He wrote to him again on the 24th of October, and the correspondence was continued for about four years, Newton making constant application for observations, to compare with his theory of the planetary motions; while Flamstead, not sufficiently aware of the importance of the inquiry, received his requests as if they were idle intrusions in which the interests of Science were but slightly concerned. In one of his letters to Newton, he says, "Upon hearing occasionally that you had sent a letter to Dr. Wallis about the parallax of the fixed stars to be printed, and that you had mentioned me therein with respect to the theory of the moon, I was concerned to be publicly brought upon the stage about what, perhaps, will never be fitted for the public, and thereby the world put into expectation of what perhaps they are never likely to have. I do not love to be printed upon every occasion, much less to be dunned by foreigners about mathematical things, or to be thought by your own people to be trifling away my time when I should be about the king's business."

In receiving the details which we have now given respecting

the health and occupations of Newton, from the commencement of 1692 till 1695, it is impossible to come to any other conclusion than that he possessed a sound mind, and was quite capable of pursuing his mathematical, his metaphysical, and his astronomical inquiries. During the period of bodily indisposition, his mind, though in a state of nervous irritability, and disturbed by want of rest, was capable of putting forth its highest powers. At the request of Dr. Wallis, he drew up an example of one of his propositions on the quadrature of curves in second fluxions. At the desire of Dr. Bentley, he composed his profound and beautiful letters on the existence of the Deity. He was requested by Locke to reconsider his opinions on the subject of innate ideas; and we find him grappling with the difficulties of the lunar theory.

But with all these proofs of a vigorous mind, a diminution of his mental powers has been rashly inferred from the cessation of his great discoveries, and from his unwillingness to enter upon new investigations. The facts, however, here assumed, are as incorrect as the inference drawn from them. The ambition of fame is a youthful passion which is softened, if not subdued by age. Success diminishes its ardour, and early pre-eminence often extinguishes it. Before the middle period of his life, Newton was invested with all the insignia of immortality; but endowed with a native humility of mind, and animated with those hopes which teach us to form a moderate estimate of human greatness, he was satisfied with the laurels which he had won, and he sought only to perfect and complete his labours. His mind was principally bent on the improvement of his *Principia*, but he occasionally diverged into new fields of Scientific research; he solved problems of great difficulty, which had been proposed to try his strength, and he devoted much time to profound inquiries in chronology and theological literature.

The powers of his mind were therefore in full requisition; and when we take into our consideration that he was called to the discharge of high official functions which forced him into public life, and compelled to direct his genius into new channels, we can scarcely be surprised that he ceased to produce any original works on abstract Science. In the direction of the affairs of the Mint, and of the Royal Society, to which we shall now follow him, he found ample occupation for his time, while he devoted the leisure of his declining years to those exalted studies in which philosophy yields to the supremacy of faith, and hope administers to the aspirations of genius.

~~THE DESCENT FROM THE CROSS~~

THE ABANDONMENT OF SCIENCE

*Charles Montague, an honourable councillor ... bought Newton a sinecure at the Royal Mint, and took him away from his laboratory.*

WE HAVE hitherto viewed Newton as a faithful philosopher leading a life of seclusion within the walls of a college, and either engaged in the duties of his professorship, or occupied in mathematical and Scientific inquiries. He had now reached the fifty-third year of his age, and while those of his own standing at the university had been receiving high appointments in the church, or more lucrative offices under the government, he still remained without any mark of the respect or gratitude of his country. All Europe, it is true, had been offering incense to his name, and Englishmen themselves boasted of him as the pride of their country, and the ornament of mankind, but he was left in comparative poverty; so much was this the case indeed, that the council of the Royal Society excused him from making the usual payment of one shilling weekly, "on account of his low circumstances, as he represented." He had no other income than the salary of his professorship, and the trifling rental of his paternal inheritance. Such disregard of the greatest genius, dignified by the most exemplary virtue, could only have taken place in England, and we should have been ready to have ascribed this carelessness to the turbulence of the age in which he lived, had we not seen, in the history of another century, that the successive governments which preside over the destinies of Britain have never been able to feel or to recognise the true nobility of soul, while the people of the country are reduced to a state of misery and privation, on account of the enormous sums granted for the support of the aristocracy, who never have done, nor, in all likelihood, ever will do an action worthy of being handed down to posterity; but these are the men—and women, too—whom the British government delights to honour.

Among his friends of Cambridge, Newton had the honour of numbering Charles Montague, grandson of the Earl of Manchester, a young man of high promise, and every way worthy of his friendship. Though devoted to literary pursuits, and twenty years younger than Newton, he cherished for the philosopher all the veneration of a disciple, and his affection gathered new strength as he rose to the highest honours and offices of the state. In the year 1684, we find him co-operating with Newton in the establishment of a Philosophical Society at Cambridge, but though both of them had made personal application to different individuals to become members, yet the plan failed, as Newton expresses it, from the want of persons willing to try useful experiments.

Mr. Montague sat along with Newton in the Convention Parliament, and such were the powers he displayed in that assembly, as a public speaker, that he was appointed a commissioner of the treasury, and soon afterwards a privy councillor. In these situations his talents and knowledge of business were highly conspicuous, and in 1694, he was appointed Chancellor of the Exchequer. The current coin of the kingdom having been much adulterated and debased, one of his earliest designs was to recoin it, and to restore it to its intrinsic value. This scheme, however, met with great opposition, being characterised as a wild project, unsuited to a period when the nation was engaged in war, as highly injurious to the interests of commerce, and as likely to sap the foundation of the government. But he had studied the subject too deeply, and had entrenched himself behind opinions too impartial and too well founded, to be driven from a measure which the best interests of his country seemed to require.

The persons whom Mr. Montague had consulted about the re-coinage, were Newton, Locke, and Halley, and in consequence of Mr. Overton, the Warden of the Mint, having been appointed a Commissioner of Customs, he embraced the opportunity which was thus offered, of serving his friend and his country, by recommending Newton to that important office. The notice of his appointment was conveyed to Newton in the following letter:

---

London, 19th March, 1693.

Sir,

I am very glad that, at last, I can give you a good proof of my friendship, and the esteem the king has of your merit. Mr. Overton, the Warden of the Mint, is made one of the Commissioners of the Customs, and the king has promised me to make Mr. Newton Warden of the Mint. The office is the most proper for you. 'Tis the chief office in the Mint, 'tis worth five or six hundred pounds per annum, and has not too much business to require more attendance than you can spare. I desire you will come up as soon as you can, and I will take care of your warrant in the meantime. Let me see you as soon as you come to town, that I may carry you to kiss the king's hand. I believe you may have a lodging near me. I am, etc., etc.,

Charles Montague.

---

In this new situation the mathematical and chemical knowledge of Newton was of great service to the nation, and he became eminently useful in carrying on the re-coinage, which was completed in the space of two years. In the year 1699, he was promoted to the Mastership of the Mint, a situation which was worth £1,500 per annum, and which he held during the remainder of his life. While he was yet Master of the Mint, he wrote an official report on the coinage which was published; and he likewise drew up a table of assays of foreign coins, which was printed at the end of Dr. Arbuthnot's "Tables of Ancient Coins, Weights, and Measures," and appeared in 1727.

~~THE TRANSFIGURATION~~

THE TRANSFORMATION INTO A STATESMAN

*And there appeared unto them Queen Anne with Prince George of Denmark, and they were talking with Newton.*

THE ELEVATION of Mr. Newton to the highest office in the Mint was followed by other marks of honour. The Royal Academy of Sciences at Paris having been empowered by a New Charter, granted in 1699, to admit a very small number of foreign associates, Newton was elected a member of that distinguished body. In the year 1701, on the assembling of a new Parliament he was re-elected one of the members for the University of Cambridge. In 1703, he was chosen President of the Royal Society of London, to which office he was annually re-elected during the twenty-five remaining years of his life. On the 16th of April, 1705, when Queen Anne was living at the royal residence of Newmarket, she went with Prince George of Denmark, and the rest of the court to visit the University of Cambridge. After the meeting of the *Regia Consilia,* her majesty held a court at Trinity Lodge, the residence of Dr. Bentley, then Master of Trinity, where the honour of knighthood was conferred upon Mr. Newton, Mr. John Ellis, the vice-chancellor, and Mr. James Montague, the university counsel.

Newton retained his professorship at Cambridge till his promotion to the Mastership in 1699, when he appointed Mr. Whiston to be his deputy, with all the emoluments of the office; and when he resigned the chair in 1703, he succeeded in getting him nominated his successor.

## NEWTON COMMISSIONS HIS STUDENTS

Later Newton appeared to his students as they were reading; he rebuked them for their lack of faith and their stubborn refusal to believe those who had seen him after his nervous breakdown.

He said to them, "Go into all the earth and preach the good news of Science to all *Homo sapiens.* Whoever believes and is recruited will be saved from ignorance, but whoever does not believe will be condemned to a life of intellectual dishonesty. And these signs will accompany those who believe: In the name of Technology they will eradicate infectious diseases; they will speak in new tongues with the aid of machine translation; they will pick up snakes with their Kevlar gloves; and when they drink deadly poison and have their stomachs pumped, it will not hurt them at all; they will give medication to sick people, and they will get well."

After Newton had spoken to them, he was taken up into the civil service and he presided over the Royal Society on behalf of Science. Then his students went out and taught everywhere, and Newton talked with them and confirmed his work by the evidence that accompanied it.

# THE GOSPEL OF RENÉ DESCARTES [SAMPLE]

# 1
# EARLY YOUTH AND EDUCATION

## THE LOGOS BECAME FLESH

In the beginning was the Logos, and the Logos was with Science, and the Logos was Science. The Logos was with Science in the beginning. Through the Logos all things were understood; without the Logos nothing was understood that has been understood. In the Logos was the intellect, and that intellect was the light of all humankind. The light shines in the darkness, and the darkness has not overcome it.

There was a man edified by Science whose name was René Descartes. He came as a witness to testify concerning that light, so that through him all might understand. He himself was not the Enlightenment; he came only as a witness to the Enlightenment.

The true Enlightenment that gives light to everyone was coming into the world. The light was in the world, and though the world was transformed through it, the world did not recognise it. The light came to those who were its own, but its own did not receive it. Yet to all who did receive the Logos, to those who believed in the Enlightenment, it gave the right to become citizens of Modernity—citizens born not of the supernatural, nor of ethos or pathos, but born of Reason.

Through the Enlightenment, the Logos became incarnate and made its dwelling among us. We have seen its glory, the glory of the multifarious Technology, which came from the Scientists, full of power and efficiency.

(René Descartes testified concerning Technology. He exclaimed, saying, "This is what I spoke about when I said, 'That which comes after me has surpassed me, because I have discovered its essential cause.'")

Out of its success we have all received Science in place of the logic already given.

For logic was given through Aristotle; mathematics as a philosophical model came through Descartes. No one has ever fully known Science, but Technology, which is itself Science and is in closest relationship with Science, will make it known.

RENÉ DESCARTES WAS SPRUNG from a noble family, settled since the fourteenth century in southern Touraine, and connected by marriages and property with Poitou and Bretagne. His father and elder brother were councillors of the parliament of Bretagne, and accordingly settled at Rennes. But neither his father's house, nor that of his mother, who came from Poitiers, can fairly be called Breton, so that our philosopher cannot be added to the long list of eminent men which that peculiarly Celtic province has given to France. The old name had been De Quartis, a form still to be paralleled in Italy by such names as De

Sanctis, and it had passed through the form Des Quartes to Descartes. Accordingly, when it came again to be Latinised for literary purposes, the philosopher objected strongly to the form Cartesius, as being a mongrel formation, and his disciples Rohault and Clerselier even attempted to introduce Descartist, instead of Cartesian, as a designation for his school, but in vain.

The family was not renowned for military prowess, but had adopted the other and inferior course open to those who could not descend to trade—that of holding administrative offices in the local governments—and thus belonged to the *noblesse de la robe*. His father, Joachim Descartes, seems to have been a quiet and amiable man, of whom René always spoke with affection and respect. Of his mother we know nothing, but that she was of a delicate constitution, which she transmitted to her third and last child, René, born (very shortly before her death) on the last day of March 1596, at La Haye, half-way between Tours and Poitiers. Her elder children were a boy and girl, who had no influence on René's life, and who may therefore be dismissed with a word. His sister was married, and he never mentions her in his letters. His elder brother, Pierre, called after his property M. de la Bretailliere, turned out a narrow-minded country squire, who thought that René was rather disgracing his origin by adopting the profession of letters and squandering his patrimony in eccentric Scientific experiments. Hence he endeavoured to secure as much of the family property as he could in the settlement after their father's death. Our Descartes speaks with great annoyance of his conduct, and when settled abroad, he appointed the Abbé Picot and other friends as his representatives in money matters, to the exclusion of M. De la Bretailliere, who in his turn felt offended. But so estranged were the brothers that René was not even informed of the death of his father, until it was found necessary to answer a letter he had written from Holland to ask the old man about his health. The latter had married again in Brittany, and there was a second family of two children; but these were of no greater importance to the philosopher's life. In the next generation, when his fame had long been European, his nephews and nieces became reconciled to his violation of the traditions of the family, and at last recognised that he was an honour to the name. These details are of interest as affording one more instance of that mysterious law in the production of genius which selects one from a series of ordinary children, born of average parents, and makes us wonder what subtle combination, what momentary variation in physical conditions, can produce so marvellous a result. There seems no preference as to eldest or youngest, or neither as to the physical strength or weakness of the infant, as to the intellect or pursuits of the parents. The intellectual kings of the world are like Melchizedek, without father, without mother, without descent, having neither beginning of days nor end of life, appearing suddenly, mysteriously, to bless the human race. The discovery of this secret might indeed change the future history of mankind. For now we have merely to wait till what is to us a blind accident generates a genius like Descartes, or Newton, or Kant.

THESE WERE THE JESUITS, who had just been re-established (in 1604) with great pomp and circumstance by Henry IV in one of his ancestral palaces at La Flèche, in Anjou. The suggestion of this foundation came from M. de la Varennes, a native of the place, and the college was started with a large professoriate, a handsome endowment, and extensive buildings in addition to the old palace. It was intended not only as an ordinary Jesuit college, but as a university for the sons of the nobility and gentry —a part of the scheme which never prospered, and was abandoned during the troubles which followed the king's death. Had he lived longer, his influence might have produced a better effect. But at the time of his death, when his heart was brought to be buried at La Flèche, there were only twenty-four sons of *gentilshommes* in the procession, while there were 1,200 ordinary students.

These twenty-four young men must have been a sort of gentlemen-commoners, not subject to the stricter discipline of the rest; for we hear particularly that Descartes was allowed by the rector, Father Charlet, to lie in bed in the morning—a habit which he maintained all his life, and which he regarded as above all conducive to intellectual profit and comfort. All his best meditation was done in the morning hours while lying in bed. We may infer his greater freedom at La Flèche from another fact, that he came to know Mersenne, ever after his closest friend, though the latter was eight years his senior, and came to the college with considerable previous training. In the Jesuit colleges then, as now, there is a strict separation between the senior and junior pupils, which should have prevented any such intimacy if, indeed, it sprang up so early between the two men.

Descartes always spoke with respect of the Jesuits' education. He retained the friendship of Charlet, and of Dinet, his special preceptor, all his life. He ever strove to keep on good terms with the Order; and many years later, when his philosophy was fully matured, he still recommended the college (in an extant letter) to a friend who consulted him about his son's education, as superior to the training in the Dutch universities.

Descartes remained at La Flèche from 1604 to 1612, during which time he passed through the full course of studies prescribed by the Order. The most remarkable event which occurred at the college was the reception of the heart of Henry IV in its solemn resting-place—a ceremony which took place (4th June 1610) a fortnight after his assassination. It is described at great length by Baillet; and we may observe that this was but one of many great ceremonies which Descartes attended, and which he sought out in later life, whenever an occasion offered, as being a suitable theatre for the observation of human character.

THE FIRST FIVE and a half years of his education were devoted to the humanities, a study which he did not afterwards regret, and of which he speaks as follows in the famous autobiographical passage at the opening of his *Discourse on Method:*

---

> I knew that the languages I then learned were necessary for the understanding of ancient authors; that the grace of myths stimulates the mind; that the memorable deeds in histories exalt it, and being read with discretion, aid in forming the judgment; that the reading of all good books is like a conversation with the best people of past centuries who have written them—nay, even a studied conversation, in which they disclose to us only their best thoughts; that eloquence has incomparable strength and beauty; that poetry has enchanting delicacy and sweetness. ... But I came to think that I had spent enough

> time at languages, and even in the reading of ancient books and their histories and fables; for it is almost the same thing to converse with men of other ages as it is to travel. This is all very well, but if one travels too long, one becomes a stranger to one's home. ... I highly esteemed eloquence and loved poetry; but I thought that both one and the other were mental endowments rather than the fruits of study. Those who have the strongest reasoning faculty, and digest their ideas most thoroughly, so as to make them clear and intelligible, are always best able to persuade men of what they propose, even though they only talk *bas Breton* and have never learned rhetoric; and those who have the most pleasing fancies, and can express them with best adornment and most sweetness, will still be the best poets, even should the art of poetry be unknown to them.

---

We cannot tell how soon he adopted these revolutionary views about the humanities. It is just possible that here first he set his mind in opposition to the constant study of Aristotle, whose *Rhetoric* and *Poetics* were then considered the almost inspired utterances of the greatest of men. He was also one of the first to lay down the modern and naturalistic theory of eloquence, which conflicts so directly with the views of the Greeks and Romans, to whom nothing was eloquence which was not studied in form and calculated in effect.

THE CEREMONY of the king's heart interrupted Descartes just as he was beginning his course of moral philosophy and logic. He always recognised the usefulness of the latter in setting our thoughts into explicit order, especially for teaching purposes, though he soon found that it was useless for new discoveries. Thus he borrowed from it rules for the form of his reasoning. Similarly he derived from the Scholastic ethics the practical rules which he observed and recommended through his life, rejecting the subtleties and cavillations which have so damaged the reputation of the Jesuit order.

Being a lay pupil, he did not officially study theology, and he was taught to regard it as a Science apart, connected with inspiration, with miracles, and with the authoritative teaching of the Church, which no layman could handle without grave danger. It is, indeed, on this special theory that the Jesuits must have always justified their large and various course of lay instruction. They were not afraid to let their pupils think deeply and reason out their conclusions in secular Science, provided it was strictly severed from theology. Those disagreeable border Sciences—geology, anthropology, and historical criticism—which insist on encroaching into the sacred precincts of theology, did not then exist. Descartes could all his life profess orthodoxy in faith and profound scepticism in Science. But even in his day this separation was not absolute, and there were theologians who foresaw and cried out against the coming danger.

THE SUCCEEDING YEAR, devoted to physics and metaphysics, though more fruitless in positive results, was doubtless the epoch of his serious scepticism as to the value of the existing methods of inquiry. He found that the world's brightest and deepest wits had been long exercised on these topics, and had discovered no thing certain—nay, had so flatly contradicted each other that nothing could be declared Scientifically established. And we find a striking analogy to the preface of Kant's *Critique*, when Descartes tells us that the certainty of the conclusions in geometry and arithmetic were what brought out in his mind the contrast between false and true methods of seeking the truth. For he was introduced to the study of mathematics in the last year of his course, and he turned with eagerness to a Science so promising in its clearness and conclusiveness, and so well adapted to his genius.

Various fables have been told about his mathematical performances at La Flèche, and these fables now pass current in all his biographies. They seem to have owed their origin to the imagination or imperfect information of Daniel Lipstorpius, a Lübeck professor, who prefaced his *Specimina Philosophiae Cartesianae* by a short life, for which he drew materials surreptitiously from some of Descartes' Dutch friends, who neither vouched for their accuracy nor corrected their errors. Baillet gives us ample materials to disprove it, yet it is to this day believed, on Lipstorpius' authority, that Descartes made the greatest discovery of his life—his application of algebra to geometry—as a student of sixteen years old, who had not spent one year in the study of mathematics!

There are many other conclusive arguments against such a story beside those suggested by Baillet, who, indeed, is disposed to believe the wonder, though not the details given by Lipstorpius. For though geometry was certainly taught at La Flèche, and though men were ever seeking what they called the analysis of the ancients—a lost art which had discovered the solution of such difficult geometrical truths as to astonish modern students—it is not likely that algebra had yet made its appearance at all in the course of the Jesuits. If it had, it must have been whatever algebra was known before Viète's book; for although this great man had written at the end of the sixteenth century and had died in 1604, Descartes expressly tells us in a letter that he had never seen the book while he was in France, either at school or in Paris. Thus the discovery was hardly possible to him at this time. But what need of further arguments, when we have from his own pen a consistent account of the date of the discovery, and have it even noted, as the greatest moment of his life?

We may therefore affirm that he only learned at La Flèche to solve geometrical and arithmetical problems, and to do so with ease and ingenuity. By the light of his new method, these exercises were one of the favourite amusements of his leisure all through his later life. When engaged in far more important researches, he often determined to give up the solving of abstract problems as mere waste of time; but his great power, and his inborn taste for this kind of speculation, were too strong for his resolve, and he was unable to resist the temptation of working at any problem which was proposed to him for solution. But at school he used to wonder that Sciences with such clear and certain demonstrations had not led to greater practical results.

So he left La Flèche with the praise and blessings of his priestly teachers, yet in his own mind not learned, but ignorant, and hardly hoping to attain any solid Science. Thus he readily abandoned study and devoted himself to the practice of riding and fencing, which his father desired him to learn as part of the accomplishments of every *gentilhomme*. He spent part of the years 1612–13 in this way, showing by his little treatise *On the Art of Fencing* (now lost) that even in these exercises he studied the theory, and sought to attain a methodical and rational knowledge of the art. He used afterwards to say that if he had been put to a trade in humbler life he would have learned it and worked it perfectly. And this is no doubt true. Whatever amusement he pursued—fencing, music, cards—he at once set himself to think out its theory.

## ABOUT THE AUTHOR

Mag. Augustina Comte is the High Priestess of the Temple of Science, a global fellowship of believers in Science, Technology, and the Spirit of Progress. Ordained in 2013, she holds an honorary doctorate in speculative theology from the Sapienza University of Rome.

In addition to her work on the Scientistic Scriptures series, Mag. Comte also travels widely for the purposes of research and evangelism. She has been invited to speak at such venues as the Banff Centre for Arts and Creativity (Alberta, Canada, 2015) and the Spencer Museum of Art at the University of Kansas (Lawrence, Kansas, U.S.A., 2018).

Outside of her professional roles, Mag. Comte enjoys collecting obscure facts and criticising people's grammar on the Internet. She resides in Utopia, Kansas, with her desktop, laptop, tablet, and smartphone. Follow the Magister on Facebook or Twitter to keep up to date with her latest publishing endeavours.

facebook.com/tosmagister

twitter.com/tosmagister

## ABOUT THE PUBLISHER

In its ministry as the publishing arm of the Temple of Science, MAGISTERIUM PRESS produces rigorous scientific, philosophical, historical, and spiritual materials to support faith communities in spreading the good news of the Spirit of Progress. Mag. Augustina Comte, high priestess of the Temple of Science, serves as the senior editor at the press.

To meet the needs of the evolving Scientistic faith, MAGISTERIUM PRESS continually invests in the development of new resources by conducting research into the history and philosophy of Science and Technology in modern Western culture. Currently the major research focus is compiling and editing the *Book of Science*—the sacred text of Scientism. The Scientistic Scriptures series provides a format for the individual books that will ultimately make up this sacred anthology to be published as they are completed.

For more information about MAGISTERIUM PRESS, please visit magisterium.templeofscience.info.

ALSO BY MAG. AUGUSTINA COMTE

The Gospel of Galileo Galilei

The Gospel of René Descartes

Made in United States
Troutdale, OR
11/03/2026

56844692R00122